Originary Technicity: The Theory of Technology from Marx to Derrida

Also by Arthur Bradley

THE NEW ATHEIST NOVEL: Fiction, Philosophy and Polemic after 9/11 (*co-authored with Andrew Tate, 2010*)

THE MESSIANIC NOW: Religion, Philosophy, Culture (*co-edited with Paul Fletcher, 2010*)

THE POLITICS TO COME: Power, Modernity and the Messianic (*co-edited with Paul Fletcher, 2010*)

DERRIDA'S 'OF GRAMMATOLOGY': A Philosophical Guide (*2008*)

NEGATIVE THEOLOGY AND MODERN FRENCH PHILOSOPHY (*2004*)

Originary Technicity: The Theory of Technology from Marx to Derrida

Arthur Bradley
Lancaster University, UK

palgrave macmillan

First published 2011 by
PALGRAVE MACMILLAN

Palgrave Macmillan in the UK is an imprint of Macmillan Publishers Limited, registered in England, company number 785998, of Houndmills, Basingstoke, Hampshire RG21 6XS.

Palgrave Macmillan in the US is a division of St Martin's Press LLC, 175 Fifth Avenue, New York, NY 10010.

Palgrave Macmillan is the global academic imprint of the above companies and has companies and representatives throughout the world.

Palgrave® and Macmillan® are registered trademarks in the United States, the United Kingdom, Europe and other countries.

ISBN 978–0–230–57692–6 hardback

This book is printed on paper suitable for recycling and made from fully managed and sustained forest sources. Logging, pulping and manufacturing processes are expected to conform to the environmental regulations of the country of origin.

A catalogue record for this book is available from the British Library.

Library of Congress Cataloging-in-Publication Data
Bradley, Arthur.
Originary technicity : the theory of technology from Marx to Derrida / Arthur Bradley.
p. cm.
Includes bibliographical references (p.) and index.
ISBN 978–0–230–57692–6 (alk.paper)
1. Techne (Philosophy) 2. Continental philosophy. I. Title.
B105.T43B73 2011
190—dc22 2011006605

10 9 8 7 6 5 4 3 2 1
20 19 18 17 16 15 14 13 12 11

Transferred to Digital Printing in 2012

Contents

Acknowledgements

First of all, I am very grateful to the friends and colleagues who helped shape my thinking in different ways throughout the process of writing this book whether by reading chapters, contributing ideas or offering general advice and encouragement: Louis Armand, Richard Beardsworth, Fred Botting, Michael Dillon, Andy Dawson, Stephen Dougherty, the late Paul Fletcher, Michael Greaney, Joanna Hodge, William Large, Laurent Milesi and John Schad.

I am also grateful to audiences at the University of Aberystwyth, the University of Cardiff, Liverpool Hope University, Manchester Metropolitan University, Roehampton University, the University of St Andrews, the University of Glasgow, the University of Groningen, the American University of Beirut, Volda University College, Norway and Charles University, Prague for listening to, and offering valuable feedback upon, earlier versions of some of the work presented here. In the process of seeing the manuscript through to publication, I have been greatly helped by Priyanka Gibbons and the editorial team at Palgrave Macmillan together with the insights of my anonymous readers.

My gratitude also goes to a number of publishers for permission to reproduce material here. Firstly, an abridged version of Chapter 5 was originally published as 'The Deconstruction of Christianity: From the Hand of God to the Hand of Man', in *Sacred Worlds: Literature, Religion and the Imagination* ed. by Mark Knight and Louise Lee (London: Continuum, 2009). In addition, I have also re-published a few paragraphs throughout the book from the following essays: '*Deus ex Machina*: Religion and Technology in Aristotle, Heidegger and Derrida', *Comparative Critical Studies*, 2: 1 (2005); 'Derrida's God: A Genealogy of the Theological Turn', *Paragraph*, 29: 3 (2006) and 'Originary Technicity? Technology and Anthropology' in Louis Armand and Arthur Bradley eds, *Technicity* (Prague: Charles University Press, 2007).

This book was begun at the University of Lancaster, UK and largely completed during a period as a visiting professor at the American University of Beirut (AUB), Lebanon. I am very grateful to Professor Samir Khalaf of the Centre for Behavioral Studies at AUB for granting me a research position and to my parents-in-law, Salwa and Hussein Hamdar, and my sister-in-law, Sarah Hamdar, for so generously hosting me during my stay.

Finally, I would like to thank my wife, Abir, for her love and support throughout the process of writing this book: I could not have begun – let alone finished – the book without it.

1
Life

At the origin there is technics (*la technique*).[1]

In the beginning, it was already a machine. It lay there, at the edge of the water, under the light of the sun. Absorbing the sun's rays into itself – trapping and storing their energy – it began to use that energy to convert the carbon dioxide in the water into an organic compound: sugar. To put it in more anthropomorphic terms, it was beginning to create its own food and, thus, ensuring that it would continue to live. By photosynthesising the gas and the water into biological matter, it also released a waste product – oxygen – into an atmosphere still largely composed of carbon dioxide and so, as yet, unable to sustain life. Such, it would appear, was the process, millions of years ago, that gradually created the conditions for more complex forms of life to develop: algae, plants, animals and, finally, humans.

It is now generally accepted that this familiar story of the evolution of life is also, in one sense or another, a story about the evolution of technology. After the Second World War, the emergence of the new sciences of molecular biology, cybernetics and information theory came together to bring about a remarkable transformation in our understanding of what the biologist François Jacob famously called '*La logique du vivant*': what we call 'life' is, in a sense that apparently goes to the very limits of analogy or metaphor, 'technical'.[2] Not only was the transmission of life (even its random elements) governed by invariant rules of reproduction, combination and adaptation – what has famously become known as the genetic code, script or programme – but this code can itself be deciphered, broken down and reassembled differently. For Jacques Monod, whose research with François Jacob and André Lwoff into genetic transmission was awarded the Nobel Prize in 1965, it was even possible to claim that all 'living beings are chemical machines'.[3] Perhaps most remarkably of all, post-war information theory arrived at a very similar conclusion at the same time albeit by a different route: Norbert Weiner and his fellow cyberneticists articulated an entirely new theory of information whose systems applied to the living and the non-living, the organic and the inorganic, the animal and the machine, alike.[4] If any system that is capable of remembering and processing information, of regulating its own behaviour and adapting to its environment deserves the name of 'technology', the philosopher Jean-François Lyotard argues in the wake of the information revolution, then even infusoria – the tiny algae synthesised by light at the edge of tidepools millions of years ago that we mentioned in the opening paragraph – are already 'technical devices'. In Lyotard's words, 'the living cell, and the organism with its origins, are already *tekhnai*': ' "life", as they say, is already technique'.[5]

What exactly does it mean, though, to say that 'life' is always already 'technique'? Is technicity simply a seductive analogy, a heuristic fiction, an increasingly well-worn cliché? Or does it describe a more profound identity between the living and the non-living, nature and technology, the gene and the machine? It is this story – the story of the emergence of what Jacques Derrida paradoxically calls the 'originary technicity' of life itself – that this book seeks to tell. As many critics have observed, we are currently in the midst of a technological 'turn' in contemporary continental philosophy: Jacques Derrida; Jacques Lacan; Michel Foucault; Jean-François Lyotard; Gilles Deleuze and Felix Guattari; and Giorgio Agamben (for all the real and often irreconcilable differences between the work of such canonical thinkers) are increasingly recognised as important and influential philosophers of technology whose work is informed by, and engages in, the new scientific revolutions of the post-war era. To offer a brief overview, Jacques Lacan is one of the first philosophical thinkers to grapple with the implications of the new cybernetic revolution after the Second World War: the psychoanalyst famously deploys the cybernetic circuit as a conceptual model for what he sees as the symbolic structure of subjectivity.[6] Gilles Deleuze and Felix Guattari's idea of a 'machinic phylum' at the heart of morphogenesis also draws on both molecular biology and post-cybernetic theory to explain the self-organising material flux of the universe as it emerges from chaos into order.[7] By the same token, the later work of Jean-François Lyotard appeals to catastrophe theory in order to articulate his theory of an inhuman remainder at the core of all humanisms (*The Inhuman*, pp. 1–7). Perhaps even the later Foucault and, differently, Giorgio Agamben's theory of bio-power – where governmental technologies are exerted upon the bare fact of life itself – can be seen as a critique of the political exploitation of the becoming-technical of the living in the aftermath of the new sciences.[8] Such a technological turn has, if anything, gathered momentum in recent years with the appearance of important works by such figures as Friedrich Kittler, Bernard Stiegler and Jean-Luc Nancy, amongst many others.[9] In what follows, however, our primary critical focus will remain on Derrida's deconstruction of the western metaphysics of presence from the perspective of an originary technicity that inhabits the interiority of life itself: 'life is a process of self-replacement', Derrida asserts, 'the handing-down of life is a *mechanike*, a form of technics' ('Nietzsche and the Machine', p. 248). This book seeks to offer a critical genealogy of Derrida's theory of originary technicity – a 'life' if you like – from its origins in a critique of Greek metaphysics through its present implications (and limits) to its future possibilities.

1

In a paradoxical sense, the technological turn in continental philosophy is as old as philosophy itself. It could not but occur because, as David Wills recently argues, what it describes is 'the turn into a technology that was always there' even from the very beginning.[10] At the outset of western metaphysics, for example, Plato establishes a famous opposition between two different kinds of memory in the *Meno*: *anamnēsis* and *hypomnēsis*. To Socrates, true contemplation or *thēoria* is nothing other than the immortal soul recollecting what it always already knows: 'searching and learning as a whole are recollection [*anamnēsis*]'.[11] However, this genuine act of memory or *anamnēsis* must be distinguished from another, artificial or technical, type of memory called *hypomnēsis*. For Plato, as he makes clear in the *Phaedrus*, external aids or supplements to memory such as the written word do not provoke recollection, so much as *forgetfulness*. On the one side, then, we have thought, the infinite, the immortal, the transcendental and something called 'philosophy'. On the other side, however, we have artifice, finitude, death, the empirical and something called 'technology'. If *anamnēsis* is living and innate, then it has no need for the sophistical supplement to memory that is *hypomnēsis*: technology is thus consigned into the darkness of the unthought. Such a foundational myth, as we will see throughout this book, is the beginning not only of western metaphysics – with its defining oppositions between the infinite versus the finite, the soul versus the body, the ideal versus the material – but perhaps also of something that endures into the modern critique of metaphysics: every attempt to oppose a pure, primary and technics-free time to a secondary, corrupt, technological time – whether it takes the form of being versus becoming (Nietzsche), clock-time versus *la durée* (Bergson), vulgar countable time versus ontological time (Heidegger) or homogeneous chronological time versus messianic time (Benjamin, Levinas, Derrida) – arguably recalls, in one way or another, the Platonic myth of recollection.

It is clear, however, that the status of *hypomnēsis* – and thus of technology itself – cannot be so easily resolved. As Derrida reminds us in 'Plato's Pharmacy', his classic reading of the *Phaedrus*, the living memory of *anamnēsis* is (like all living things) finite: what needs to be recollected had, after all, first to be forgotten. For Derrida, then, the key question is – if *anamnēsis* is unable to retain everything all by itself – then what helps it to do so, and his answer is, of course, *hypomnetic* supplements: tales, notes, chronicles, memorials and, finally, writing. Perhaps one might also recall here that even Plato proves his theory of *anamnēsis* by virtue of a hypomnetic technique: Socrates draws geometrical shapes in the sand with a stick for Meno's slave who then innately 'remembers' what they represent. If Plato's so-called living, self-present memory *always* requires this technical, non-living form of memory to help it recall the past, Derrida argues, then *hypomnēsis* can no longer be seen as a mere supplement or aide-memoire that is entirely exterior to the philosophical logos: 'Memory is ... contaminated with its first substitute: *hypomnēsis*.'[12] Just as Derrida sees technology as always already at work in memory itself, so his former pupil Bernard Stiegler arrives at a similar conclusion from the opposite direction: what we call living memory is always already in the process of *exteriorising itself* onto non-living technical supplements. In Stiegler's later account, the Platonic opposition between *anamnēsis* and *hypomnēsis* – together with the exclusion of technology from the sphere of memory, thought and the logos – represents nothing less than '*the denial of the originary technicity of memory*'.[13]

To be sure, what exactly Plato and Aristotle meant by the term '*technē*' (τέχνη) is a vast question that lies beyond the scope of this study.[14] It is clear, nonetheless, that one very particular classical definition of the technological object has, rightly or wrongly, persisted above all others. For Aristotle, as he makes clear time and again in the *Physics*, a technical artefact is an essentially inert, neutral tool or instrument with no capacity to move itself. On the one hand, a natural being such as an animal or a plant contains 'within itself a principle of motion and of stationariness (in respect of place, or of growth and decrease, or by way of alteration)'. On the other hand, however, a technical object such as 'a bed and a coat and anything else of that sort – in so far as they are products of art – have no innate impulse to change'.[15] If *phusis* (φύσις) innately contains the origin (*arche*) of its own motion – which is the reason why an acorn will grow into an oak tree all by itself – the same is obviously not true for a technical or fabricated artefact: an oak bed frame requires what has become known as an external efficient cause (*causa efficiens*) (though, as Martin Heidegger has famously pointed out, this term actually appears nowhere in Aristotle's discussions of technology) to bring it into being. Such, anyway, is the theory of technology that has dominated philosophy for more than 2000 years: the technical artefact is a *prosthesis* (*pro-thesis*, literally, that-which-is-placed-in-front-of) to nature, thought and the human, with no formative or reproductive power of its own, that can be utilised for good or ill depending upon who or what happens to wield it.

In a significant gesture, however, Aristotle can be seen to blur the distinction between *phusis* and *technē* just as Plato did with *anamnēsis* and *hypomnēsis*: what is originally designated a technics-free zone – nature, thought, the logos – once again turns out to be contaminated by technics. It is striking, to start with, that – for all the ontological differences between them – the philosopher constantly describes the workings of nature via an analogy or comparison with technical processes. To Martin Heidegger, whose 1939 essay 'On the Essence and Concept of φύσις in Aristotle's *Physics* B, I' charts the gradual reduction of the Pre-Socratic concept of *phusis* into the modern techno-scientific idea of 'nature', Aristotle's *phusis* has historically been boiled down to little more than a species of self-causing or making – what we would today call automatic – *technē*.[16] Such an incremental technicisation of *phusis* can already be found, for instance, in the philosopher's famous description of nature operating like a doctor doctoring himself (*Physics*, 199b). Nevertheless, one might argue that Aristotle is only able to define *phusis* in terms of a quasi-mechanical automation without fear of contradicting himself because he has already taken the decision to deprive *technē* of any possible autonomy of its own. If the technological artefact cannot move by itself (because it has no innate impulse to change), then what moves by itself cannot, by definition, be technological – and so must be deemed 'natural'. For Bernard Stiegler, what is conspicuously absent from Aristotle's circular logic here is even the bare theoretical possibility of a *self*-causing or self-organising technology (*The Fault of Epimetheus*, p. 1). Perhaps what makes this omission all the more interesting, though Stiegler does not mention this, is that

Aristotle is prepared to countenance precisely such a technology in the famous thought experiment on automation from the *Politics*: 'suppose', the philosopher imagines, 'that every tool we had could perform its task, either at our bidding or itself perceiving the need, and if – like the statues made by Daedalus or the tripods of Hephaestus, of which the poet says that "self-moved they enter the assembly of the gods" – shuttles in a loom could fly to and fro and a plucker play a lyre of their own accord, then master craftsmen would have no need of servants nor masters of slaves'.[17] What, one starts to wonder, would even the theoretical possibility of such an entirely self-moving technology mean for Aristotle's attempt to oppose *phusis* and *technē* on the basis of motion?

2

In many ways, the next key philosophical moment in the emergence of contemporary theories of originary technicity is the sixteenth and seventeenth centuries with the birth of the epoch of modern science. It is with the proper names Galileo, Descartes, Newton, Leibniz and Hobbes, in particular, and the scientific revolution in general, that we witness the first concerted attempt to establish a relation – whether analogical or literal – between nature and technics via the theory of mechanics. According to the historian of science Amos Funkenstein, the modern concept of 'science' represents the coalescence of four distinct ideals: homogeneity, univocity, mathematisation and, most pertinently from our perspective, mechanisation.[18] For Funkenstein – just as for Heidegger before him – the invention of this new discipline called 'science' also called for the invention of a new *object* of scientific study called 'nature' which is radically different from either the Greek *phusis*, the Roman *natura* or the Christian *ens creatum*: what science calls 'nature' is pure matter, passive, inert, uniform, subject to immanent mechanical laws and forces (rather than Aristotelian teleological ends) which are predictable, determinable and even manipulable by rational, scientific analysis. Such is the epistemic backdrop to the classic question of post-Galilean mechanistic philosophy: how far can any 'living' organism – the body, the mind, the soul, even the universe itself – be compared to a machine?

It is worth pondering whether the machine metaphor for organic life – for all its force – may still depend on a certain denial or forgetting of a more essential machinism within that life: the reason why the slave could be so completely replaced by an automaton in Aristotle's *Politics*, recall, is because the slave is *himself* nothing more than a kind of animated machine, a living tool (*Politics*, 1, 2, 1253b). As Georges Canguilhem points out, it was Aristotle – rather than, as commonly believed, Descartes – who was the first thinker to compare the human body to a mechanism when he draws a parallel between the operation of the body's articulated joints to devices such as the catapult and the pivot in *De Motu Animalium*.[19] For Descartes himself, the thought experiment of the human catapult is famously replaced by that of the spring-operated human clock ticking away all by itself:

> I might consider the body of a man as a kind of machine equipped with and made up of, bones, nerves, muscles, veins, blood and skin in such a way that, even if there were no mind in it, it would still perform all the same movements as it now does in those cases where movement is not under the control of the will or, consequently, of the mind.[20]

In the theatre of Cartesian dualism, of course, it is *only* the body that can be compared to the machine – the soul remains absolutely immaterial – but nonetheless from here it is clearly only a short step for a subsequent generation of materialists like Julien Offray de le Mettrie to extend the same metaphor to cover the mind as well: 'the soul is clearly an enlightened machine'.[21]

At the end of the eighteenth century, indeed, the machine metaphor for life had exceeded its materialist roots and even permeated the transcendental idealism of critical philosophy. Take, for example, Immanuel Kant's attempt to offer a natural basis for human moral freedom in the Third Critique via the device of a teleological 'technic' of nature.[22] To begin with, Kant insists that human beings must be seen as having a naturally given predisposition to free rational activity: 'Man is indeed the only being on earth that has understanding and hence an ability to set himself ends of his own choice [*willkürliche Zwecke*] and in this respect he holds the title of lord of nature' (*Critique of Judgement*, p. 318). However, it is only because man has the (natural) capacity to set himself ends that he is able to see nature as end-driven at all: what nature lends man is the very ability to see nature *as* giving in the first place. For Kant, this human capacity to see organic life as organised purposively or teleologically in terms of ends is what he calls the 'technic of nature': 'Given that we find something purposelike in nature's products, let us call nature's procedure [*Verfahren*] (causality) a technic' (p. 271). Now, the philosopher presents the technic of nature as nothing more than a metaphor – presumably because, unlike Aristotle before him, he is unable to even conceive of a genuinely autonomous, self-organising technology (p. 237) – but there are still grounds for wondering whether it is just a metaphor. Perhaps Kant is right to say that it is only our own power of teleological judgement – rather than nature itself – that is properly (*eigentlich*) technical (p. 408), but we must still bear in mind that, in the Kantian schema, this power is anything but arbitrary: it is the a priori, universal and necessary procedure by which reflective judgement unifies empirical intuitions with concepts. If we cannot but use such judgements, and we cannot but see nature as technical if we are to constitute it at as an object of experience at all, then the idea of a technic of nature becomes, perhaps not literally true, but a kind of *necessary* metaphor – the obligatory 'as if' of an Idea of Reason. In Pheng Cheah's Derridaean reading, we might go still further: what the necessity of thinking nature technically entails is not the sheer impossibility of thinking nature 'in itself' but a kind of original lack *within* the natural that calls for such technical supplementation.[23]

For the post-Romantic materialists of the mid-nineteenth century – including, most notably, the historical materialism of Karl Marx – we can detect a new engine driving the machine metaphor for nature: the emerging science of thermodynamics. It is not too much of an exaggeration to say that thermodynamics was to the nineteenth century what mechanics was for the seventeenth and eighteenth, and information theory would become for the later twentieth and twenty-first – quite simply the basic conceptual, political and even ethical organising ground of life itself. As Anson Rabinbach argues in his cultural history of thermodynamics, *The Human Motor* (1992), the German scientist Hermann von Helmhotz's discovery of the principle of the conservation of energy finally levelled post-Hegelian *Naturphilosophie*'s lingering speculative opposition between matter and spirit or nature and culture. To Helmhotz's eyes, *all* forces of nature – whether animal, human or even machinic – are simply forms of a single universal energy (*Kraft*) and subject to the immutable laws of conservation and entropy that govern it.[24] Perhaps most importantly, as the title of Rabinbach's study indicates, the new science of thermodynamics not only revolutionised the study of nature – setting the seal upon the mechanistic turn of early modern science – but of human being and, particularly, human labour: we can no longer identify anything qualitatively singular about our mode of Being-in-the-World to distinguish it from that of animals or machines. If Aristotle could speak of the human catapult, and Descartes of the human clock, nineteenth-century philosophical anthropology (as Jacques Lacan would observe 100 years later) adopts a new analogy: the human motor or engine operating by a strict calculus of input/output ratios. Such is apparently the basis not only for the Marxian theory of labour but also for the Freudian model of the psyche as the site of self-regulating drives like the Pleasure Principle: '[i]t dawned on people, something which was never thought of before, that living things look after themselves all on their own, in other words, they represent homeostats', Lacan writes (*Seminar*, Book II, p. 75). In a gesture that would have enormous ramifications into the twentieth century, the human body becomes just one more thermodynamic energy-producing machine amongst others (*The Human Motor*, p. 77).

What provisional observations can we make on the basis of this very rough schematic history of the relationship between *phusis* and *technē*, nature and technics, the organic and the mechanical? It is worth noting that, from Plato up to Descartes, the traffic between the two only ever flows in one direction: technics is consistently employed – whether positively or negatively – as a heuristic device for articulating what, if anything, is proper to life. As Canguilhem points out in his study of the machine metaphor, the organism is always compared to a preconceived and often very limited idea of the machine – lifeless, mechanical, prosthetic – rather than the machine being compared to the living, autonomous and self-forming organism ('Machine and Organism', p. 45). Yet, to pose the question in this thoroughly anthropological way – 'to what extent am I like a catapult or a clock or a steam engine?' – is, of course, almost always to presuppose a certain kind of anthropological answer. To risk a somewhat speculative hypothesis, I wish to suggest that metaphysical philosophy of technology effectively becomes an ironic, self-fulfilling *mechanism* – an 'anthropological machine' to borrow Giorgio Agamben's influential term – designed to produce, define and shape not simply 'technology' but, more importantly, *what is not* technology: '*Homo sapiens*, then, is neither a clearly defined species nor a substance', Agamben writes, 'it is, rather, a machine or device for producing the recognition of the human'.[25] On the one hand, philosophy posits a strategically impoverished theory of technology: lifeless, inert and prosthetic. On the other hand, it valorises a metaphysical natural or human essence that absolutely exceeds technicisation: *anamnēsis*, *phusis*, the soul, the *ego cogito*. Such a gesture, as we will see in the chapters that follow, logically and necessarily excludes an empirico-transcendental middle ground where *phusis* and *technē*, the living and the non-living, mechanism and organism, come together. If ancient philosophy's theory of technology is obviously and inevitably limited by the historical condition of technology itself – for Plato and Aristotle could hardly be expected to have any conception of an autopoietic or self-organising technology – such an excluded middle cannot simply be explained historically, because what is most remarkable about the thinkers we have examined is the extent to which they are still able to confront *theoretically* the possibility of a more general technicity within life itself, even or especially in the act of denying it: what is Aristotle's slave, after all, if not a species of what we would today call artificial intelligence? Perhaps, in this light, we might see the subsequent materialist turn in modern philosophy of technology – which increasingly sees no qualitative difference between *phusis* and *technē* whether as mechanisms or productive forces – as a return of this repressed originary technicity. In nineteenth-century philosophy of technology, what Aristotle saw as a strictly limited analogy between organism and mechanism is effectively transformed into something close to a tautology.

3

In later nineteenth-century and twentieth-century thought, however, philosophy of technology undergoes another series of epistemic historical, ontological and metaphysical leaps that create the space for the contemporary technological turn. It goes without saying, of course, that this philosophical revolution rhymed with – and was arguably even outstripped by – a technological revolution as well: the epoch of techno-science moves from the apogee of industrialisation to the birth of the information age in little more than 100 years. After Descartes' spring-operated clock, a whole array of new heuristic machines rush to fill the philosopher's imagination: Marx's steam engines and water mills; Freud's Mystic Writing Pad; Heidegger's hydroelectric dams, power stations and allusions to quantum physics; Lacan's automatic camera; Derrida and Stiegler's digital technologies and, today, the auto-hetero-poietic machine that is the self-organising, self-regulating, adapting and emerging organism or body. To start with, Karl Marx and Friedrich Engels' historical materialism clearly constitutes a decisive inaugural moment in this history: Marx is today, as we will see in Chapter 2, widely recognised as the first modern *'penseur de la technique'* (Kostas Axelos) whose radical break with the Aristotelian theory of technology can be detected everywhere from the Frankfurt School's critique of technological instrumentalism to Hardt and Negri's embrace of a Neo-Spinozan machinic materialism. If Marx is one critical touchstone for contemporary theories of technics – in the sense that all the work in the field is in dialogue, whether silent or audible, with the challenge posed by historical materialism – Martin Heidegger represents another: what the later Heidegger calls the 'question concerning technology [*die Frage nach der Technik*]' under the figure of *das Gestell* is taken up, problematised and re-posed by Jacques Derrida and Bernard Stiegler. Such a profound ontologisation of technics – where it essentially becomes coterminous with the disclosure of Being as such – again prepares the ground for what Derrida and, later, Stiegler will call originary technicity.

It is, of course, impossible within this brief history to enumerate all the thinkers who have contributed in different ways to the construction of the contemporary theory of technics: Henri Bergson; Edmund Husserl; Ernst Jünger; Walter Benjamin; Theodor Adorno and Max Horkheimer; Norbert Weiner; Gregory Bateson; Jacques Ellul; Bruno Latour; Maurice Merleau-Ponty; Herbert Marcuse; Gilbert Simondon; André Leroi-Gouhran; Humberto Maturana, Michel Serres, Francisco Varela and Paul Virilio are only the most distinguished names in this lineage. To single out just one key figure, Simondon's phenomenology of individuation and technological evolution is a major (if still underrated) influence not only on Deleuze's theory of multiplicity but also on Stiegler's concept of originary technicity. For Simondon, the theory of psychic and collective individuation transforms the individual subject or entity into the *effect* of a pre-individual field that both precedes and succeeds it: a process like crystallisation is a paradigm of individuation in this respect inasmuch as individual crystals do not pre-exist but emerge out of a pre-individuated metastable environment. Perhaps most crucially, Simondon's theory of technogenesis also represents a radical departure from the Aristotelian model of hylomorphic production: a technical object is not the result of form (*morphē*) being stamped upon matter (*hylē*) by a pre-existing efficient cause so much as the 'transduction' of a diverse set of elements coming into a singular new relation that precedes any individual human act of making.[26] In Bernard Stiegler's philosophy, as we will see later on in this book, Simondonian transduction becomes a privileged means not only for conceptualising the correlation between technics and temporalisation but of the pursuit of new psychic and collective modes of individuation in the digital epoch.

Perhaps another neglected resource for contemporary philosophy of technics whose presence will be felt throughout the pages of this book is the fascinating, if now somewhat dated, work of the palaeoanthropologist André Leroi-Gouhran. To Leroi-Gouhran's eyes, the post-war discovery of the prehominid life form Zinjanthropus – an Australopithecus that was erect, bipedal and tool-using but had, as yet, developed only very limited brain capacity – required a complete re-writing of the process of human evolution to stress the constitutive role manual technologies played in the development of human intelligence. By obtaining an upright posture, it seems that Zinjanthropus freed his hands for tool-use; tool-use, in turn, is what liberated the lower jaw for language and the lower jaw, finally, liberated the brain for symbolic thought and intelligence. If Simondon attacks hylomorphism by offering a new phenomenology of technological evolution, in other words, Leroi-Gouhran's own critique takes the form of a new, and quite literally bottom-up, anthropology. Whereas the conventional wisdom insisted that human beings evolved a larger brain (three or four times the size of that of a chimpanzee) and then used it to develop human culture, Leroi-Gouhran turns this logic on its head: it was our departure from genetically fixed patterns of behaviour that led to ever-increasing brain capacity. Such was another way in which technics became 'originary' for the human: what Zinjanthropus reveals is that defining human qualities such as consciousness, intelligence and the capacity for symbolic thought are not the cause of tool-use but an effect. For Leroi-Gouhran, moreover, the story of the human from the evolution of the skeleton all the way up to the age of the computer is the story of the progressive liberation or 'exteriorisation' of its capacities through technology: 'The whole of our evolution has been oriented towards placing outside ourselves what in the rest of the animal world is achieved inside by species adaptation.'[27] In my discussion of both Derrida and Stiegler's rival accounts of originary technicity, we will see how both thinkers take up and radicalise Leroi-Gouhran's account of life's exteriorisation.

In returning to the present day, however, we must also go back to where we started: what Derrida calls originary technicity can profitably be situated between two distinct, parallel and yet intertwined scientific revolutions – the technological and the biological, the cybernetic and the molecular, the circuit and the gene – which have transformed our understanding not simply of the human but of 'life' itself. It is important to recall here that at the very beginning of his career, Derrida himself briefly but suggestively situated the impossible project of a science of writing – a grammatology – within a larger 'scriptural turn' at work in both information theory and molecular biology which sought to transform life into a code or program: 'Now we tend to say "writing" for all that and more.'[28] To be sure, the philosopher himself makes clear that such parallels are by no means an invitation to reduce grammatology to a 'positive science' – quite the contrary – but, nonetheless, a considerable body of literature has emerged in recent years that has insightfully placed deconstruction against the backdrop of the new sciences: cybernetics, molecular biology, information theory, systems theory, quantum physics and so on.[29] By reading Derrida and his contemporaries side by side with such thinkers as Norbert Weiner, Humberto Maturana and Francisco Valera or Niklas Luhmann, it becomes possible to witness the extent to which deconstruction might contribute to such defining debates and concepts as artificial intelligence, complexity, adaptation and emergence or the embodiment, extension and distribution of mind. Consider, for example, the ways in which the theory of autopoiesis has inserted itself into the continental philosophical imaginary over the last 30 years or so. For many contemporary critics, Maturana and Valera's image of a self-making, self-regulating and self-regenerating autopoietic machine represents a kind of litmus test for the originary technicity of life:

> [It] is a machine organised (defined as a unity) as a network of processes of production (transformation and destruction) of components which: (i) through their interactions and transformations continuously regenerate and realize the network of processes (relations) that produced them; and (ii) constitute it (the machine) as a concrete unity in space in which they (the components) exist by specifying the topological domain of its realisation as such a network.[30]

Perhaps most crucially, autopoiesis recognises no qualitative difference between organic and inorganic systems: all living systems are autopoietic, and so any physical system – whether social, cultural, artificial – can, if autopoietic, be said to exhibit life (*Autopoiesis and Cognition*, p. 48). If the original concept of autopoiesis arguably remained insufficiently genetic to account for the possibility of technological evolution and adaptation as such (for what it described were essentially closed, circular and autonomous systems), Francisco Valera's later work on biological autopoiesis has gone on to open up to a more heteropoietic, fluid and dynamic process of interaction between an organism and its environment.[31] In his recent work on the relation between autopoietic theory and deconstructive philosophy, posthumanist critic Cary Wolfe has highlighted an intriguing parallel between the two: both articulate an aporetic point of chiasmus between inside and outside where each constitutes the other (*What is Posthumanism?*, pp. 1–30). Such a point of 'structural coupling' – where the organism evolves through meeting and interacting with its environment – is also a key, if as yet largely unrecognised, influence on what contemporary philosophers like Stiegler understand by originary technicity (*The Fault of Epimetheus*, p. 158).

4

What, then, to bring this opening chapter to a close, is at stake in Derrida's theory of originary technicity? Where does it come from? And what might be its future? It is with such questions that we come to the subject of the remainder of this book. As our provisional *tour d'horizon* has already shown, though, the answers are far from clear: we have unsurprisingly found little or no agreement over the meaning of *technē* itself, let alone in what sense (metaphysical? biological? mechanical? informational?) it could be meaningfully described as 'originary'. Yet, as a bottom line, what all the thinkers we will consider in this book would agree upon – despite the very real differences between them – is that *technē* is no longer thinkable as a mere prosthesis that is placed in front of life, nature or the human. To establish a kind of basic ground zero definition, then, originary technicity is less a tool or prosthesis that has been super-added *to* life nor even quite a metaphor *for* life but what I will call the empirico-transcendental condition *of* life itself.[32] Such an aporetic condition is articulated phenomenologically, historically and even ontologically by different thinkers under such names as labour, matter, the real, Being-in-the-World, the other and the body, but the basic gesture remains the same: what is supposedly *outside* the sphere of the human, nature and life is constantly folded back *inside* it as its 'ground'. If the classical philosophy of technology is a machine for producing the non-technological, in other words, then contemporary theories of originary technicity see themselves as a machine for revealing that technology is always already contaminating *phusis*, *anamnēsis*, consciousness, ipseity or the living more generally. On the one hand, we will see that originary technicity problematises a certain metaphysical idea of the *origin*: full, simple, natural, indivisible and plenitudinous. On the other hand, however, we will also see that originary technicity destroys the traditional idea of *technology*: inert, instrumental, an empirical or material *positum*. In one slightly perverse sense, to which we shall return at the very end of this book, we might go so far as to argue that contemporary philosophy of technics is scarcely a philosophy of 'technology' at all inasmuch as the empirico-transcendental technological 'condition' it seeks to describe is both everywhere and nowhere: technics is *not*.

It is my intention in the following chapters to offer a critical introduction to, and a genealogy of, Derrida's theory of originary technicity through a series of case studies of key figures up to, including and beyond Derrida himself: Karl Marx; Sigmund Freud and Jacques Lacan; Martin Heidegger; Jacques Derrida and his philosophical pupil Bernard

Stiegler, and, finally, the burgeoning field of posthumanist theory. Accordingly, what follows is – though it presumes no prior expertise or knowledge of the field on the part of the reader – by no means intended to be all-embracing or comprehensive in scope. To begin with, my approach inevitably remains largely deconstructive in orientation – which is not to say that I am uncritical of Derrida's philosophy nor that I am unaware that a very different book could have been written from the perspective of, say, post-Marxism, Deleuze's materialism or Badiou's mathematical ontology. If I do seek to represent many of the dominant trends within continental thought over the last 100 years – Marxian historical materialism; psychoanalysis; fundamental ontology; phenomenology; deconstruction; genealogy – my selective approach results in some important and inevitably rather arbitrary omissions as well: Nietzsche; Simondon; Deleuze/Guattari; Virilio. Perhaps most importantly, however, what follows is not simply a simple history or genealogy of Derrida's theory of originary technicity but an attempt to offer a *critique* of that theory. In this book, I wish to pursue a line of enquiry – opened by a diverse group of scholars of technics including, perhaps most aggressively, the embodiment theorist Mark B. N. Hansen[33] – that there is a significant blindspot at the heart of contemporary theories of technics.

To state it very simply, I wish to argue that the theory of originary technicity from Marx to Derrida is – despite its claims to the contrary – *not technical all the way down.* It will become clear over the course of this book that something – or rather someone – always precedes or exceeds technicisation. As we have seen, the theory of originary technicity claims to reveal the essential 'technicity' of life itself – right down to the process of replication of the biological cell – but I wish to argue that it consistently privileges *one* form of life over all the others. For me, originary technicity – whether transcendentalist or empiricist, idealist or materialist, phenomenological, humanist or posthumanist, ontological, deconstructive or genealogical – still remains in the thrall of what Agamben calls the anthropological machine: it is a mechanism for producing and recognising the being that we ourselves are. Let me briefly adumbrate the two interconnecting ways in which we will be able to witness this machine at work in the chapters that follow:

- On the one hand, originary technicity persistently focuses on the question of what technology means *for* a (retrospectively pre-technicised) human subject to the exclusion of all other forms of life, whether animal, biological or purely physical.

- On the other hand, originary technicity correspondingly forecloses the possibility of technology as an autonomous material force or dynamic which may have an evolutionary trajectory that is quite separate from the (after all comparatively provincial) sphere of human subjectification.

Finally, and most crucially, what this means is that contemporary philosophy of technics effectively ends up caught within the very anthropocentric imaginary it seeks to explode: it consistently presupposes an (embodied or disembodied, thinking or feeling, self-cognisant or auto-affective) human subject as the ultimate tribunal in front of which the claims of technology can be adjudicated *even or especially when it is that very subject whose existence is being called into question.* Such a spectral or ghostly subject – who never quite dies or, better still, seems to die *infinitely* – is thus effectively granted the position of the presiding judge at its own trial. In each of the six chapters that follow, we will see how the theory of originary technicity from Marx to Derrida risks turning itself into little more than what Timothy Clark has called a 'non-essentialist anthropology': what is intended to produce a de-anthropologising of the human – originary technicity – ironically risks becoming a new means of defining the *anthropos*.[34]

Firstly, we turn to Karl Marx and Friedrich Engels' philosophy of technology. It is Marx who represents the first major touchstone for Derrida's theory of originary technicity: historical materialism proposes a radical critique of the Aristotelian concept of *technē* as an essentially inert, neutral prosthesis that survives all the way to deconstruction. As we will see, Marx's theory of technics stretches across the entire range of his thought from the early philosophical anthropology of the *Economic and Philosophic Manuscripts of 1844* all the way up to the mature thinker's critique of the real subsumption of labour by machine stage capital in the *Grundrisse* and *Capital*. For Marx, we cannot simply *oppose* the human to technics in the way that Aristotle opposes the *causa efficiens* to the *prosthesis*: what the philosopher famously defines as the 'labour process' in the *Capital* is nothing less than an originary interface between humanity and technology whereby each invents and constitutes the other across time. If Marx's insight into the mutual co-constitution of the human and the technical is undoubtedly a radical step, however, we will go on to consider whether his political critique of the machine (in which a collective human subject can emancipate itself from domination by technology and regain control over what should be its tool) re-capitulates an implicitly Aristotelian concept of technology-as-prothesis. To what extent does Marx – and even contemporary Marxian critics like Hardt and Negri – reinstate Aristotle's efficient cause in collective historical and dialectical form under the figure of the proletariat?

Sigmund Freud and Jacques Lacan's psychoanalytic theory of technology is the topic of Chapter 3. It is now well documented that Freud proposes a series of intriguing analogies between the operation of the psyche and various technical or machinic systems all the way from the early *Project for a Scientific Psychology* to 'A Note on the "Mystic Writing Pad"'. To recall Freud's own deeply ambiguous claim, our psychic mechanism closely resembles a species of writing. Yet, according to a number of commentators – most notably Derrida and Lacan – Freud is unwilling to push the analogy between mind and machine to its logical conclusion: the unconscious might *literally* be a species of writing machine. For Lacan – whose 1954–55 seminars famously claim to go back to the materialism of early Freudian psychoanalysis – it remains entirely possible to articulate the very 'materialist definition of consciousness' that Freud systematically refuses: consciousness is *indeed* a kind of symbolic machine – a ticking clock, a steam engine, an automatic camera, a cybernetic circuit or feedback loop – that operates, as it were, 'all by itself'. If Lacan's little-known thought experiment extends Freud's insight into the psyche-as-machine still further, however, this chapter will consider whether we can once again detect a residual anthropocentrism even in this radical critique of the Cartesian *ego cogito*: Lacan's focus still remains upon the instrumental role that machines play in revealing the symbolic status of human activity. To what degree does Lacan's radicalised version of the mind/machine metaphor still reduce the machine to a mere supplement to the process of psychic subject-constitution?

Martin Heidegger's ontological *destruktion* of *technē* is the subject of Chapter 4. It is now almost universally acknowledged that Heidegger is – for better or worse – the single most important and influential twentieth-century philosopher of technology, and his influence can be felt in everything from deconstruction to cognitive science. As is well known, his questioning of technology stretches from the analysis of equipmentality (*das Zeug*) in *Being and Time* before culminating in the pivotal essays of the 1940s. For the later Heidegger, of course, the 'essence' (*Wesen*) of technology is famously nothing to do with ontic or empirical technology itself so much as with what he calls the *Gestell* that constitutes the dominant revelation of Being in the epoch of technoscience: everything – including humanity itself – is revealed as a 'standing reserve' (*Bestand*) of energy to be liberated and stockpiled.

If Heidegger consistently seeks to maintain the ontological priority of *das Gestell* over any ontic or empirical technology, this chapter will consider whether his philosophy merely raises the Aristotelian reading of technology to a new ontological level: the Heideggerian poet or thinker now plays the role of the efficient cause whose pious questioning enables the technical to come into presence. To what extent does Heidegger's ontology return empirical technics to the same inert state of instrumental passivity to which Aristotle originally consigned it in the *Physics*?

Jacques Derrida's own philosophy of originary technicity will be assessed in Chapter 5. It is no exaggeration to say that the whole of deconstruction – from his early work on the *arche*-writing that precedes all language all the way up to his late reflections upon the virtual or artefactual status of space and time itself – can be read as a radical philosophy of technics. To put it in Derrida's own words, what is called 'deconstruction' is nothing other than the deconstruction of the historic opposition *between* thought and technics. Such a re-working of the empirico-transcendental divide lies at the basis of his critique of a residual anthropologism in Freud, Husserl, Heidegger and even Jean-Luc Nancy's self-proclaimed 'Deconstruction of Christianity'. If Derrida thus seeks to articulate the technicity of thought itself, we will see that his philosophy has nonetheless been subject to charges that it, too, contains a residual idealism: the deconstruction of the opposition between thought and technology is apparently capable of being expedited by thought alone. For Stiegler, amongst others, Derrida's version of originary technicity remains a largely formal and phenomenological proposition which – like Heidegger's *Gestell* – privileges the intentionality of an implicit (if problematised) *ego cogito* and occludes the constitutive role that ontic technical supplements have historically played in the formation of thinking. To what degree does Derrida's philosophy of technics perpetuate the reduction of technics to a question to be posed by, about and for an implied thinking subject?

Bernard Stiegler's increasingly influential philosophy of technics will be examined in Chapter 6. It is the major hypothesis of Stiegler's multi-volume *Technics and Time* that there is an essential relation between the human process of temporalisation and our mnemo-technical supplements. As we saw in our discussion of Plato's myth of *anamnēsis*, however, Stiegler insists that the western metaphysical philosophy of time transforms this essential relation into an opposition between a pure, primary and technics-free time and a secondary, supplemental technical time. For Stiegler, as he goes on to show in a wide-ranging empirico-transcendental analysis of the relation between the human and technology, we can no longer oppose technics and time: all temporality is constituted technically. Perhaps most powerfully, Stiegler extends this philosophical claim into a political critique of what he calls the hyper-industrialisation of time under contemporary capital: the human capacity to experience time – itself constituted technically – is now subject to a gesture of real subsumption by the mnemo-technical industries. If Stiegler's philosophy appears to offer the most uncompromising expression of originary technicity within contemporary thought, however, we will go on to consider whether – like Derrida before him – he entirely manages to dispel the spectre of a pre-technical *ego cogito* who remains capable of thinking the technological condition of its own time consciousness. To what extent does even Stiegler's version of originary technicity end up rehearsing the metaphysical opposition between technics and time?

Finally, Chapter 7 turns to the future to survey the field of posthumanist, transhuman and even *inhuman* theory. To risk stating the obvious, posthumanist theory is a broad and complex field that encompasses everything from an almost technophiliac millenarianism at one extreme to a kind of neo-Romantic organicism or naturalism at the other. On the one side, we have neo-transcendentalists like Hans Moravec for whom technogenesis will ultimately realise the Cartesian dream of a disembodied, post-organic virtual consciousness. On the other side, we have embodiment theorists like Katherine N. Hayles and Mark B. N. Hansen, who insist that the organic body remains the indispensable biological and phenomenological locus for any process of technogenesis. Yet, in spite of such differences, I will ask whether it still remains possible to detect a residual anthropocentrism – both explicit and implicit – across the entire spectrum of posthumanist theory. If embodiment theory is right to attack the Cartesian dualism at work in the desire to simply transcend the body, we will consider whether the theory of embodiment itself remains implicitly Cartesian: what embodiment calls the 'body' ends up performing an uncannily similar role – self-constituting, auto-affective, self-reflexive – to what Descartes calls 'mind'. Perhaps a more intriguing history of our posthumanist future, to conclude, might lie in a new genre of philosophical anthropology that has emerged over the last decade or so where the story of the human race is narrated from the perspective of the (at least allegedly) *inhuman*: solar death (Jean-François Lyotard); a robot historian, rock formations, germs and viruses (Manuel de Landa) or species extinction (Quentin Meillassoux). To what extent do such anti-human thought experiments finally offer a genuinely post-anthropological theory of originary technicity – or is some ghostly human subject still summoned in order to preside over its own technological demise?

5

In *Originary Technicity: The Theory of Technology from Marx to Derrida*, then, I seek to offer one version of the past, present and future of the technological turn within continental philosophy. It is important to bear in mind from the outset, however, that, just as this turn was always already there, so it is by no means 'over' – indeed, if what I have been claiming is true, it can never end. To offer one last *caveat lector* before concluding this opening chapter, then, I should stress that what unites the theories of technics adumbrated in this book is scarcely a common conceptual, philosophical or even methodological agenda than an increasingly vertiginous set of *questions*. What does it mean to speak of an originary technicity? When – if ever – did such a theory come into being? Where will it go in the future? How does it change the way we think about life (*zoe* and *bios*), nature (*phusis*, *natura*, biology) and the being that we ourselves are (*Zoon politikon*; *Homo sapiens* and *faber*; *ego cogito*; *Dasein*; gene carrier; posthuman cyborg)? And *who* – if anyone – will turn out to be the thinker – the *res cogitans* – of that thought?

2
Labour

> Labour is the eternal natural condition of human existence [*Arbeit ist ewige Naturbedingung menschlischer Existenz*].[1]

In the view of a number of influential commentators from Kostas Axelos up to Jacques Derrida himself, Karl Marx is nothing less than the first thinker of technology – *le premier penseur de la technique*.[2] It is with Marx, in other words, that the philosophy of technology is deemed to move decisively beyond the Aristotelian instrumentalist account of *technē* that had dominated thought within the western metaphysical tradition. As we have just seen in Chapter 1, Aristotle's *Physics* see '*technē*' [τέχνη] as nothing more than a '*prosthesis*' – a tool or instrument – that requires what has become known as an efficient cause (*causa efficiens*) to set it in motion: 'a bed and a coat and anything else of that sort – in so far as they are products of art [*technē*] – have no innate impulse to change' (*Physics*, 192b–193b). However, it is precisely this ancient idea of technics as an inert, neutral instrument deployed for some end that pre-exists it, that historical materialism seeks to challenge, from the philosophical anthropologies of the early Marx to the mature thinker's critique of industrial capital. Such is why Marx's philosophy of technology is the starting point for our own genealogy of originary technicity.

For Marx, we can no longer simply oppose humanity and technology as if they were entirely separate ontological entities: each (according to a process that will have to be defined carefully) comes into existence through, and as, the other. If Marxian philosophy is almost universally acknowledged today to represent a break with the Aristotelian instrumentalist account of technology, however, it is revealing that it is also routinely convicted of not going quite far *enough*: Bernard Stiegler, for example, argues that in historical materialism 'technics finally continues to be thought as a means [*moyen*] of production, to the extent where the metaphysical understanding of time, which is not questioned, still dominates that philosophy of technics' (*Les Temps du cinema*, pp. 135–6n). Perhaps the most famous index of Marx's alleged inability to think of technics as anything other than an Aristotelian means to a non-technical end is his critique of the machine stage of capital in the *Grundrisse* whereby an apparently pre-technological collective human subject can obtain freedom from the alienation incurred by the formal subsumption of its labour under industrialisation. In this way, the machine is restored to its (rightful?) place as the instrument of man's production of himself and his world.

This chapter explores Marx's philosophy of technology from the anthropology articulated in early works like the *Economic and Philosophic Manuscripts* to the theory of labour propounded in the *Capital* before concluding with the famous critique of machines in the *Grundrisse*. To what extent is the *premier penseur de la technique* also the first thinker of what Derrida calls originary technicity?

1

In the opening pages of his *La Philosophie de Marx* (1993), Etienne Balibar offers the following salutary *caveat lector*: 'there is not and there never will be a Marxist philosophy'.[3] One might add that the same holds true for Marxist philosophy of technology. As Alfred Schmidt observes, Marx's theory of the relationship between man, nature and technology arises out of an exceptionally diverse and mixed set of sources – from German idealism through political economy to scientific materialism – and itself evolves across the philosopher's career.[4] First of all, the young Marx's philosophy of technology remains recognisably Hegelian in outline: his theory of labour in the *Manuscripts* presupposes a concept of a unique human essence or 'spirit' seeking to actualise itself by shaping the otherwise inert matter of 'nature' in its own image. Yet, if this residual humanism never entirely disappears in the later Marx, a number of critics – most famously, of course, Louis Althusser – have argued that it is gradually (or abruptly) superseded by a new scientific or technological materialism. For Althusser, the later Marx saw the labour process – not as the expression of a neo-Hegelian human essence – so much as of a 'material mechanism' that is dominated by 'the physical laws of nature and technology'.[5] Perhaps the most important factor driving Marx's so-called 'epistemological break', Anson Rabinbach and (to a lesser extent) Amy E. Wendling argue, is the emerging theory of thermodynamics which finally destroyed any speculative opposition between man and nature by seeing *all* forces of nature – whether natural, human or technological – as forms of a single universal energy or *Kraft*.[6] If the young Marx remains committed to a theory of labour as the expression of human spirit in contradistinction to nature, Rabinbach goes so far to argue that the mature Marx sees no qualitative difference whatsoever between human labour and any other inorganic productive force in nature or technology: the human body is just one more thermodynamic energy-producing machine in the universe (*The Human Motor*, p. 77). In Bernard Stiegler's contribution to the reception history – which I wish to tease out at greater length in what follows – Marx/Engels' theory of labour moves beyond both the man-versus-machine humanism of speculative idealism and the man-as-machine reductivism of thermodynamic theory by articulating a new and dynamic structural coupling between the human and the tool which serves to organise both (*The Fault of Epimetheus*, p. 2).

It is with the young Marx's theory of labour, though, that we will begin. According to Derrida, of course, the founder of historical 'materialism' was not essentially a thinker of being as matter at all: 'He is a thinker of being as work.'[7] Despite being thoroughly immersed in the new science of the 1840s – Ludwig Feuerbach, Jacob Moleschott, Ludwig Büchner et al. – the philosopher remained sceptical about the kind of 'vulgar' materialism that sought to mechanise and thus de-historicise the relation between man and nature: labour is what puts the 'history' into historical materialism (*The Concept of Nature in Marx*, pp. 86, 92). To the Marx of the 1844 Paris *Manuscripts*, for instance, it is already axiomatic that labour is not something added to human life or being so much as the basic condition of that existence: 'Life itself appears only as a *means to life*.'[8] For the young philosopher, man is undoubtedly a 'living natural being' endowed with vital powers and instincts but – precisely because he *is* a natural being – he is also a 'suffering, conditioned and limited creature' who is utterly dependent on external objects to maintain his existence:

> That is to say, the *objects* of his instincts exist outside him, as *objects* independent of him; yet these objects are *objects* that he *needs* – essential *objects*, indispensable to the manifestation and confirmation of his essential powers. (*Economic and Philosophic Manuscripts*, p. 181, Marx's emphases)

By changing 'inorganic nature' in accordance with his needs – turning natural resources into food, heating, clothes and a dwelling place – the natural world becomes the objectification (*Vergegenständlichung*) of man's own powers: it is nothing less than man's '*inorganic body*' (p. 112). Such (self-) productive life – in which man actualises or externalises his own nature *onto* the inorganic world and recognises himself within, and as, that world – is the basis of what the young Marx famously (or notoriously) calls human species-existence (*Gattungswesen*): '[i]t is just in his work upon the objective world, therefore, that man first really proves himself to be *a species being*' (p. 114). In this way, the young Marx effectively anthropologises Hegel's philosophy of nature: human labour becomes the means by which 'spirit should find its own essence, its counterpart [*Gegenbild*], i.e., the Notion [*Begriff*], within nature'.[9]

As Marx's philosophy develops through the 1850s, however, the theory of labour becomes (for a variety of possible reasons upon which there is still remarkably little agreement) increasingly de-humanised: what begins as the neo-Hegelian actualisation of the human essence through nature turns out – by the time of Chapter 7 of the *Capital* – to be no longer a singularly or uniquely human process. What are the symptoms of this famous epistemological break? To start with, it is striking that the mature Marx now defines the labour process as a *triangular* relationship between man, nature and one other (hitherto obscure) element:

> The simple elements of the labour process are (1) purposeful activity, that is work itself (2) the object on which that work is performed and (3), the instruments of that work. (*Capital*, p. 284)

If the first two factors in that process – the vital needs of man and the resources of nature which will fulfil them – are already familiar to us from the Paris *Manuscripts*, the introduction of the instrument of labour as the third element in the definition of labour is important here: labour cannot proceed without the instruments of labour. For Marx, that is, the instrument represents the essential point of interface between the labour of man and the subject of his labour without which no labour can take place:

> An instrument of labour is a thing, or complex of things, which the labourer interposes between himself and the object of his labour, and which serves as a conductor, directing his activity onto that object. (*Capital*, p. 285)

Perhaps we can already begin to detect Marx's break with the Aristotelian instrumentalist account of technology here: what the philosopher calls the instrument of labour is no mere prosthesis but the materialised medium (*The Concept of Nature in Marx*, p. 103) by which the dialectic of man and nature can take place. In Marx's account, moreover, the instrument – as the means through which man's self-actualisation necessarily conducts itself – ultimately becomes as essential to the sustenance of his life as his own bodily organs: 'just as a man requires lungs to breathe with, so he requires something that is the work of human hands in order to consume the forces of nature productively' (*Capital*, p. 508).

Yet, if we now have an abstract definition of the instrument of labour, we are no nearer to understanding what historically and materially such an instrument might look like, and Marx's examples are revealing. Firstly, and most briefly, Marx floats the important idea (to which we will return later) that man's own *body* – his hands and feet – is itself an instrument: 'man's bodily organs alone serve as the instruments of his labour' when gathering such 'ready-made means of subsistence' as fruits (p. 116). Afterwards, though, the philosopher quickly puts this idea to one side in favour of the theory – first raised in the *Manuscripts* – that the natural world forms a kind of inorganic organ or instrument: nature itself is man's 'original tool house', he argues, providing him with 'stones for throwing, grinding, pressing, cutting, etc' (p. 117). For Marx, however, it is ultimately neither man's own body nor the natural world that defines the human labour process but something else:

> The use and construction of instruments of labour, although present in the germ among certain species of animals, is characteristic of the specifically human labour-process, and Franklin therefore defines man as a 'tool-making animal'. (p. 117)[10]

Just as labour is the natural condition of human existence, so this natural condition expresses itself through the use of instruments and, almost from the very start, through non-natural or fabricated tools. If technological implements are an elementary factor in the labour process, and the labour process is, in turn, allegedly the defining characteristic of human being, then Marx is able to conclude with Benjamin Franklin that human species-existence is defined through its relation to technology: man is a tool-using animal. In this sense, we might be tempted to conclude that Marx is not simply a thinker of being as work (as Derrida argues) but of being as technics: the *premier penseur de la technique* is also a thinker of technics as first philosophy.

To be sure, Marx's definition of man as *homo faber* is a famous and important step in his historical materialist anthropology, but we are still only half way towards understanding what happens in the labour process. It is not quite enough to say that man actualises his own vital essence or spirit by using tools to transform the natural world, because Marx is very clear that man has no essence *before* this process of actualisation. According to a logic that anticipates Derrida radicalisation of Leroi-Gouhran's theory of exteriorisation (as we will see in Chapter 5), the becoming-actual of the essence through the use of tools is what retroactively *constitutes* that essence *qua* essence in the first place. Not only does Marx's anthropology consistently take the man/nature or spirit/matter opposition to its rhetorical limit here but what characterises the human body in his account at times seems to anticipate the kind of 'extended' or 'distributed' quality – both projecting itself outwards into its inorganic environment, on the one hand, and incorporating that environment into itself as its own artificial limbs, on the other – that contemporary cognitive philosophy now regularly attributes to mind.[11] For Marx, as the philosopher goes on to make clear in a famous passage from Chapter 7 of the *Capital*, labour is itself nothing other than this process of organico-inorganic coupling at work:

> Labour is, first of all, a process between man and nature, a process by which man, through his own actions, mediates, regulates, and controls the metabolism [*Stoffwechsel*] between himself and Nature. He confronts the materials of nature as a force of nature. He sets in motion arms, legs, heads and hands, in order to appropriate the materials of nature in a form adapted to his own needs. Through this movement he acts upon external nature and changes it, and in this way he simultaneously changes his own nature [*Indem er durch diese Bewegung auf die nature außer ihm wirkt und sie verändert, verändert er zugleich seine eigne Natur*]. He develops the potentialities slumbering within nature, and subjects the play of its forces to his own sovereign power. (*Capital*, p. 115)

What is being described in the dialectic of the labour process? Firstly, it is important to observe that – since the alleged humanism of the *Manuscripts* diagnosed so influentially by Althusser – the position of labour in Marx's analytic has undergone something of a structural transformation: 'labour' *itself*, not the labourer, has become the in-human material subject that converts man and nature into one another. If 'labour' merely named the point of articulation between the otherwise autonomous substances of man and nature – spirit and matter – in the young Marx of the *Manuscripts*, the mature Marx of *Capital* seems to use it to name a more dynamic, recursive and mutually constitutive exchange of forces between organism and environment.[12] Perhaps the most important index of labour's constitutive role in this relationship can be found in Marx's claim that *both* man *and* nature are transformed through their mutual participation in the labour process: what starts out with man transforming nature in accordance with his own needs by using tools, ends up with man's own nature itself being transformed. By changing nature with tools, man changes his own nature, in other words, but what I want to insist upon in what follows is the *ontological* (rather than merely prosthetic or instrumental) role that technology performs in this theory of labour: man the maker *of* tools is at the same time made *by* those tools.

For Marx, to recapitulate, human species-existence (if we can still use such a term to name something in which the human itself is only one element) is both *productive of*, and *produced by*, its technical interactions with its environment: man and matter invent one another through the medium of the tool. It becomes possible, in the light of such a conclusion, to imagine a Marxian materialist genealogy (rather than a simple anthropology) of the human where, as Amy Wendling notes, 'the embodiment of different forms of tools produces different types of human being' (*Marx on Technology and Alienation*, p. 140). As Marx himself puts it in the *Grundrisse*, 'Not only do the objective conditions change in the act of reproduction' – as wastelands become cultivated fields and villages towns – 'but the producers change, too, in that they bring out new qualities in themselves, develop themselves in production, transform themselves, develop new powers and ideas, new modes of discourse, new needs and new language.'[13] What exactly, though, are these 'new qualities'? Firstly, labour exerts itself upon the labourer on a physical – which is to say bodily – level: labour does not merely mobilise the natural forces of the body – 'arms and legs, heads and hands' – but transforms its physical form, power and capacity. By beginning to labour manually in the first place, the human body expends 'brains, muscles, nerves, hands, etc' (*Capital*, p. 137) and, as the labour process evolves, that body progressively develops becoming mentally and physically stronger, more knowledgeable and so on. However, the physiological transformation of the body through the labour process can also be witnessed negatively: whereas manual labour and even the manufacturing stage of industry train and strengthen the body the machine stage of capital notoriously emasculates, feminises and infantilises the male labouring body (§15, pp. 492–564). Perhaps the classic example of how the labour process itself constitutes human being is, however, not to be found in the body at all but in consciousness: the mind, as much as the arm or the leg, is a product of labour. If Marx starts out from the neo-Aristotelian position that human species-being makes and uses tools in order to fulfil preconceived ends – this is the species difference between human labour, on the one hand, and the labour of bees or spiders, on the other (pp. 283–4) – he ends up arguing that the tool, in turn, generates new ends – new needs, new instincts, new ideas – that did not pre-exist its use: 'the satisfaction of the first need, *the action of satisfying, and the instrument of satisfaction which has been acquired,* leads to new needs' (emphasis added).[14] The preconceived idea and the imagined end of human activity are generated by, and through, an *inhuman* instrument of satisfaction. This is why Marxian philosophy of technology breaks so radically with the hylomorphism of Aristotelian theory of technology. In Marx's account, what is apparently most natural to human existence – the creation, production and maintenance of its life – is at least partially devolved to the sphere of the technical: the idea of the tool is in the tool itself.

Perhaps we can flesh out the Marxian account of the mutually constitutive relation between human and tool by turning to Friedrich Engels' fascinating thought experiment 'The Part Played by Labour in the Transition from Ape to Man' in the *Dialectics of Nature*.[15] It is Engels' objective from the very outset of this piece to radicalise Marx's account of labour as the 'natural condition of human existence' to the point where it does not simply *describe* our basic mode of species-existence – we are what we do – but *constitute* that existence in the first place: 'we have to say that labour created man himself' (*Dialectics of Nature*, p. 279). According to the notoriously problematic trajectory of the *Dialectics* – which seek to reconcile the twin 'sciences' of dialecticism and evolutionary biology into a thoroughgoing natural materialism – we must understand this claim about labour's role in forming the human species in Darwinian, almost Lamarckian, terms: socio-cultural evolution runs alongside, and ultimately overtakes, biological evolution. Despite the fact that much of the *Dialectics* belongs to the realm of historical curiosity today, it is remarkable that Engels' (almost entirely speculative) thesis on the origins of man still manages to anticipate many of the theories of modern palaeo-anthropology, particularly Leroi-Gouhran's argument about the role of tool use in human evolution.[16] To start with, Engels argues that the birth of the human race can be traced to the moment when prehominid life attained an upright posture and thus became able to use its upper limbs for new purposes: '*the hand became free* and could henceforth attain ever greater dexterity and skill' (p. 281).[17] What exactly, though, makes the human hand so handy? For Engels, the answer to this question is, unsurprisingly, tool-manufacture and use: 'The specialisation of the hand... implies *the tool*, and the tool implies specific human activity, the transforming reaction of man on nature, production' (p. 17). If everything begins with the liberation of the hand for tool-use, then the evolution of the brain – of the cerebral cortex, of consciousness, of the capacity for symbolic thought – necessarily comes later: what we call mind is the after-effect of the sociability that manual labour immediately entails (pp. 284–5). Perhaps most importantly, however, Engels agrees with Marx that the transformer is *himself* physically transformed through his use of technology – and nowhere is this more obvious than in the case of the hand. One the one side, the hand is the 'organ of labour': it is the meta-tool – a five digit gripper with opposable thumbs – that enables man to use tools in the first place. On the other side, though, the hand is also '*the product of labour*': it has itself evolved and adapted as an organ through a long process of prehominid labour activity. In Engels' account, the hand is thus the product of what (anticipating Stiegler in Chapter 6 of this book) we might call a *co-evolution* of human and tool, organic and inorganic matter (*The Fault of Epimetheus*, p. 137):

> Only by labour, by adaptation to ever new operations, by inheritance of the resulting special development of muscles, ligaments, and, over longer periods of time, bones as well, and by the ever-renewed employment of these inherited improvements in new, and more complicated operations, has the human hand attained the high degree of perfection that has enabled it to conjure into being the pictures of Raphael, the statues of Thorwaldsen, the music of Paganini. (p. 281)

What provisional conclusions can we draw from Marx's historical materialist anthropology about his philosophy of technology more generally? It is well known that the philosopher's corpus is littered with a bewildering number of references – ontological, anthropological, historical, political and purely speculative – to something called 'technology', but what is much more difficult to locate is a consistent position or stance upon its meaning. As a cursory reading of the history of

Marx criticism reveals, the philosopher has routinely been depicted as everything from a thoroughgoing technological determinist to an anti-technological humanist and – in the case of Althusser – even both at different stages of his career.[18] Yet, as Keith Ansell Pearson has noted, what makes this debate so sterile today is that the different positions – technological or humanist – are often just two sides of the same anthropocentric coin: either the collective human subject retains its position as the true agent and *telos* of history or historical agency is devolved purely and seamlessly onto technics.[19] For Marxian philosophy of technology, what arguably remains to be thought within his theory of labour are not the false dualisms of humanism or technocentrism, agency or determinism, cause or effect and so on but rather what we have called the mutually constituting 'between' of the human-tool relation. On the one hand, we clearly have an organic body – but an organism that only exists by exteriorising itself *onto* its inorganic environment, adapting that environment to its needs. On the other hand, we undoubtedly have an inorganic environment – but an environment that continually insinuates itself into that body, inventing the forms human species-being will take. To understand an originary human/technical relation as the motor force of history is to explode the simplistic ontology of a single autonomous agent – whether you call it 'man' or 'technics' makes no substantive difference – that is both origin and end of history and replace it with a more complex and dynamic field of force. If Marx's own philosophy does not always do conceptual justice to this force field – we will see in the second half of this chapter how his political critique of capital risks *re-ontologising* the human over and against the technical – it does begin to map a hitherto undetermined feedback loop between organism and environment that will be explored further by future theorists of originary technicity from Simondon to Stiegler. In each case, we move from a model in which the human and its milieu exist as autonomous entities or regions – and only enter into a relation through the blunt instrument of hylomorphic production – towards a *heteropoietic* system of mutual exchange.

2

In Marx's theory of labour, then, the philosopher articulates a new materialist anthropology: human species-existence is the site of a dynamic interface between the organic and the inorganic out of which the form of the human is produced. It is thus very clear that – contra the critique of figures like Stiegler (*Les Temps du cinema*, pp. 135–6n) – technics cannot be restricted to the metaphysical category of pure means of production in the Marxian text: labour is nothing less than the condition of human consciousness and this necessarily includes, as we will see, human time consciousness. Accordingly, I want to pursue this insight into the originary technicity of the human to its end by exploring how it transforms the other major dimension of Marx's philosophy of technology, namely, his critique of industrialised or machine stage capital. To be sure, Marx's story of the progressive alienation (*Entäusserung*) of human labour under capital to the point where the proletariat becomes dominated by the very technology it used to control is a familiar – indeed far too familiar – tale. Perhaps this is why it has also proved to be an easy (and again perhaps too easy) target for contemporary continental philosophy of technology: Timothy Clark, for example, argues that Marx's theory of alienation presupposes the idea of an 'untrammelled human essence' that still remains within the orbit of the Aristotelian instrumentalist theory of technology ('Deconstruction and Technology', p. 249). For Bernard Stiegler, though, it is now incumbent upon us to re-interpret Marx's critique of machine stage capital in light of the fact that the labouring subject's fall into alienation has always already happened: mechanisation comes *after* humanity's own act of self-exteriorisation onto technical artefacts.[20] If it seems pointless to deny that Marx does indeed appeal at key moments throughout his thought to the idea of a free, collective and unalienated human labour – which once had technology under its control and will do so again – I nevertheless want to wager that this concept of 'an untrammelled human essence' which has been lost and may be regained cannot, strictly speaking, be found *anywhere* in Marx's own work after the Paris *Manuscripts*. Such is the extent to which the first thinker of technology might be read as the original thinker of originary technicity. What happens, then, when we re-visit the Marxian critique of the machine in the light of his theory of the technological condition of the human itself?

It is important to begin by clarifying exactly what Marx *means* by the 'machine' in the industrial phase of capital because it differs from influential classical accounts by such figures as Pascal or Babbage. As he makes clear in the *Grundrisse*, Marx's theory is predicated on Babbage's division of machinery into two categories – those that transmit a pre-existing force like man power or natural power and those that produce their own power (p. 690) – but, of the two, it is always the *productive* machine that most interests Marx. First of all, the Marxian machine has autonomy: what distinguishes it from the handheld tool or instrument is that technology has finally become liberated from the dependence on human activity that characterised the feudal mode of production.

During the Industrial Revolution, man ceases to be the maker or controller of technology and simply becomes what Marx calls its 'motive power' – the force required to initially set it in motion. Yet, as the process of industrialisation continues apace, even this reduced role is eliminated as more powerful natural forces like wind, water and steam take man's place as the motor of technology (*Capital*, p. 497). Now, even this state of affairs was not satisfactory because natural forces are not powerful enough to produce consistent and continuous motion (the wind does not always blow, the water levels rise and fall) so a new and more dependable motive force must be secured. For Marx, the process of technological autonomisation only reaches its logical conclusion when technology has become independent enough to be its *own* prime move or motive force: the birth of the age of the machine takes place when machines – and the paradigmatic example for Marx is of course Watts' steam engine – begin to power themselves and other machines (p. 506).[21] Secondly, and following on from this, Marx argues that what makes the machine even more powerful is its systematicity of production: an individual machine is already an ensemble performing a number of operations simultaneously (p. 497). If there is one machine, in other words, then this always implies the existence of *other* machines and ultimately of machine systems, all working together in tandem, like the factory. Finally, and most notoriously, Marx also defines the machine as the logical conclusion of the capitalisation of production: the machine embodies the real subsumption of living labour – labour that has already been formally subsumed through the imposition of the wage economy – by dead or frozen labour. The species of alienation synonymous with the machine age actually pre-dates machine technology – the manufacturing workshops that dominated production from the sixteenth century through to the end of the eighteenth had already transformed the labourer into an automatic implement (p. 458) – but from there is it is only a small step to replace the human worker with a machine tool or system of tools that can perform the same job much more cheaply and productively. This is accomplished via: (1) the extension of the working day to accommodate the increased productivity of the machine in comparison to human labour; (2) the appropriation of women and children into the labour process at the expense of men as the demand for muscular strength is made redundant by the machines; (3) the driving down of wages as skilled labour is made redundant and the consequent creation of a vast surplus labour army and, finally, (4) the ever-increasing intensification of what human labour remains in an unvarying, monotonous labour process (pp. 229–58).

In this machinic subsumption of all the labourer's physical and mental powers, the tool-making animal described in Chapter 7 of the *Capital* is finally transformed into the machine-operated tool of Chapter 15:

> In handicrafts and manufacture, the workman makes use of a tool; in the factory, the machine makes use of him. There the movements of the instruments of labour proceed from him, here it is the movements of the machine that he must follow. In manufacture the workers are the parts of a living mechanism. In the factory we have a lifeless mechanism which is independent of the workers, who are incorporated into it as its living appendage. (p. 548; see also *Grundrisse*, pp. 692–3)

However – if the historical materialist anthropology set out in the first section of this chapter holds any force at all – things clearly cannot be so simple. To pose the obvious question, what *exactly* – if anything – is being alienated in the machine stage of capital? It cannot be some universal human essence that now passes into the machine because Marx makes very clear in Chapter 7 of *Capital* that the human is nothing outside its dynamic and recursive relation to its inhuman environment: our essence is non-essential from the very start. On the contrary, Marx would entirely agree with Leroi-Gouhran and later Stiegler that the human lives through a process of self-objectification: we ek-ist by putting ourselves (our consciousness, our know-how and our limbs) outside ourselves and into mnemo-technical repositories that are never simply prostheses. Any attempt to posit a theory of technological alienation as the deprivation of some primordial mental or physical capacity thus merely recapitulates the Platonic critique of artificial memory (*hypomnēsis*) from the perspective of recollection (*anamnēsis*). For Marxian philosophers of technology, then, any appeal to an originary human essence that is only subsequently alienated by the advent of the age of the machine must be put into abeyance: Kostas Axelos comes closer to my own position here in speaking of a state of 'originary alienation' – a process of essential self-exteriorisation or prostheticisation – that constitutes the 'human' itself (*Marx, penseur de la technique*). If Axelos is right to insist that humanity has always existed in a state of alienation due to our use of technology, though, it is necessary to radicalise his (still too humanist and existentialist) claim that such originary alienation is only a necessary dialectical moment on the path to the realisation of the full human essence under communism: what we must persist in calling 'alienation' – though exteriorisation might be a better word here – is less a dialectical moment to be sublated than an ontological condition to be lived. Perhaps more daringly, this means that we must also join Stiegler in re-reading the critique of the machine stage of capital – less as the alienation of some primordial essence–than as the logical *outworking* of this state of originary self-alienation: it is precisely the progressive objectification or exteriorisation of knowledge, memory and gesture into artefacts that makes the real subsumption of labour (and even of the social if Antonio Negri is correct) possible *because whoever owns or controls such artefacts necessarily also controls the 'essence' or species-existence of the human itself.*[22] On the one hand, the technicisation of the human body can be seen as inherent in the use of even the most primitive tools: the apparently simple act of using a flint axe, as Leroi-Gouhran insists, is the result of a very complex process of acquiring the capacity of anticipation and retention through repetition. On the other hand, the process of industrial capitalisation must be viewed as the *literalisation* of this motor capacity within human labour: modern industry is, as Stiegler argues, the generalisation and intensification of this 'mnemo-technological reproductive capacity of the motor behaviour of *producers*'.[23] In the same way, Jennifer Bajorek concludes that the industrial capitalisation of labour is a logical corollary of the 'infinitely extensible or prostheticisable productive power that was *already there* in labour'.[24] Such a claim by no means disqualifies a Marxian critique of industrial capital – originary exteriorisation or prostheticisation takes place in different historical forms, containing different modalities and distributions of power, and is capable of being economised upon differently – but it does forbid any appeal to an unalienated human labour as the privileged site of that critique.

Perhaps the most revealing expression of the tension at the heart of Marx's theory of alienation can be witnessed in his (hitherto somewhat obscure) philosophy of time. As we have already seen in Chapter 1, the metaphysical opposition between thought and technics is originally a temporal one: Platonic *anamnēsis* (recollection) versus *hypomnēsis* (artificial memory). Yet, the modern critique of metaphysics does not so much suspend this opposition as recapitulate it differently in the form of a series of oppositions between a pre-technical 'pure' temporality and a 'contaminated' technical temporality: clock-time versus duration (Bergson); vulgar historical time versus ontological time (Heidegger); historical time versus messianic time (Benjamin, Levinas, Derrida, Agamben). To many eyes, Marx's own philosophy of time is no exception to this rule: Peter Osborne has shown in an important article how the former's thought presupposes an opposition between the abstract quantitative time of capital and the absolute, qualitative time of human potentiality.[25] On the one side, for example, we have the chronological clock-time that is uniformly imposed upon labour from outside to measure value, productivity, the length of the working day and so on (*Capital*, Chapter 10).[26] On the other side, though, we have the immanent time of unalienated labour whose creative becoming cannot be measured according to any predetermined external scale (*Grundrisse*, p. 488). For Osborne, Marx's time of human becoming even goes so far as to anticipate Heidegger's ontological time insofar as it reveals existential finitude to be the hidden basis of the labour theory of value: 'human finitude, in the sense of the existential register of mortality, is the *ontological basis of the "value" of time*', Osborne argues, 'and thereby, the ontological ground of labour-time's functioning as a universal measure of value' ('Marx and the Philosophy of Time', p. 20). Yet, one might wonder whether we can oppose ontological time to the clock-time of capital quite so easily here or, to put it differently, whether even the time of immanence is not also a form of exteriorisation. If Marx does indeed prefigure Heidegger's phenomenology of finitude, in other words, it might be possible to level the same objection to him that Stiegler (as we will see in Chapter 6) levels against Heidegger in the first volume of *Technics and Time*: *Dasein*'s finitude – its possibility of inheriting the past into which it has been thrown and of resolutely projecting it forward into future *ekstases* – is itself only possible *because* that past has always already been inscribed for it in mnemo-technical support systems.[27] Just as what Heidegger calls ontological time can only be opposed to chronological time via a metaphysical reduction – the calendar, the act of writing and even the diurnal rising and setting of the sun are all 'clocks' *avant la lettre* (*The Fault of Epimetheus*, p. 223) – so we might wonder whether Marx's unalienated labour time might itself be understood as an earlier form of (supposedly capitalistic) chronological time: why else can the philosopher speak of something called the 'normal working day' – as opposed to the working day imposed by capital which stretches into the night in its pursuit of surplus value – if not a naturalised but universally operative sense of clock-time? Such a position – whatever its ultimate merits – would be at least one way of resolving the tension between the two competing versions of time set up by Marx: *all* time – whether qualitative or quantitative, ontological or chronological, alienated or unalienated – is technical.

For Marx himself, of course, it is in Notebooks 6 and 7 of the *Grundrisse* – which Negri calls the 'Fragment on Machines'[28] – that we obtain his most detailed account of why the real subsumption of labour under machine capital must fail. As is now very well known, his thesis is that the inherently contradictory logic of capitalism *itself* is what produces the conditions that will bring about its destruction: 'Capital is the moving contradiction, [in] that it presses to reduce labour time to a minimum, while it posits labour time, on the other side, as sole measure and source of wealth' (*Grundrisse*, p. 706). Let us briefly re-trace the steps of this remarkable argument. Firstly, Marx holds that capitalism's constant reduction of necessary human labour time through the real subsumption of labour has the ironic effect of *maximising* free or disposable time for the individual labourer. By becoming surplus to the labour process, the labourer is thus freed to pursue his own 'artistic and scientific' development, a process which transforms him into a 'different subject' who produces differently (p. 712). Now, the second point of contradiction is that, whilst it is busy rendering human labour time superfluous via the industrialisation of production, capitalism continues to dogmatically posit such labour time as the sole measure of value. Given the fact that industrial capital mobilises techno-scientific forces precisely in order to maximise labour time *beyond* human capacity – a steam hammer obviously works much more efficiently and productively than a human hand – such a continuing reliance upon exploiting human labour as the *only* source of value compels it to delimit the very forces it unleashed in the first place. Such is the moving contradiction at work: capital 'calls to life all the powers of nature and of science to make the creation of wealth independent... of the labour time employed on it', on the one hand, but 'confine[s]' the 'giant social forces thereby created' within 'the limits required to maintain the already created value as value', on the other (p. 706). Perhaps one easy way of seeing capitalism's residually anthropocentric theory of technology – which insists that technics must still be measured by the very human valence it is designed to exceed – are those moments in industrial history when capitalism has (contrary to its own ideology) operated as an effective *break* upon technological innovations because they threaten a monopoly of a particular market: witness its continuing investment in horse power and railways over the automobile in the late nineteenth and early twentieth centuries, for example, or its unenthusiastic support of green or eco-friendly technologies at the beginning of the twenty-first century. If Marx himself was notoriously vague about what exactly the communist future will look like – but how could a historical materialist be anything else when the material conditions that would enable us to imagine it have not yet arrived? – what *is* clear is that it will be a technological future. The conflict between man and machine under the condition of real subsumption is a false one created by the capitalist mode of production (*Capital*, pp. 554–5) and it will disappear. This means that the communist state will be *more*, not less, technologically advanced than its predecessor: technics, socially organised, collectively owned and released from the handbrake of value, will maximise its productivity in the interests of the whole species. In the rare glimpses Marx affords us of the communist future, it is one in which humanity finally assumes the collective subject position of technological mastery: man is able to devolve necessary labour almost entirely to the machine (*Grundrisse*, p. 705) and devote his energies to the 'true realm of freedom' which is 'an end in itself'.[29]

What, though, does the philosophy of originary technicity signify for Marx's political critique of machine capital? Is it simply impossible to sustain the theory of the alienation and re-appropriation of labour if there *never was* and *can never be* an unalienated human labouring subject? Or does it place the burden upon us to imagine new and less mutually destructive ways in which the human and technics can continue to invent one another? To risk stating the blindingly obvious, Marx's prediction of the demise of industrial capitalism in the Fragment on Machines has not come true – but what is perhaps more revealing about his philosophy of technology is why he thought it could in the first place. It is only with the philosophical failure of the Marxian critique that its historical finitude becomes visible, like an object left behind on the beach as the tide withdraws. After 100 years of the globalisation of capital, the idea that there might still exist a remainder (let alone a surplus) of disposable time for the labourer to develop his or her own means of artistic and scientific production has become incredible even for Marxists: culture, as Theodor Adorno and Max Horkheimer told us, is every bit as industrialised as cotton or steel. For Michael Hardt and Antonio Negri, by the same token, we are no longer merely living through the real subsumption of *labour* – a process to which there is still theoretically an 'outside' – but of the *entire* social, cultural and biological sphere: capital relations have expanded to 'subsume all aspects of social production and reproduction, the entire realm of life'.[30] Perhaps the most important index of Marx's own commitment to the idea that there is and can be an 'outside' of capital – a mode of subjectivity which capitalism cannot economise upon at the heart of its very drive to economy – is, of course, that this external limit becomes the historical and ontological ground of communism: the outside of the old system expands to become the inside of the new system. Just as the capitalist industrialisation of production ironically generates more and more disposable time for the labourer, so this time is re-appropriated by the collective under communism: the socialised subject – whose labour is replaced by the machine – is set free to pursue his own development. On the one hand, the human – emancipated from labour by technology – is re-installed in the neo-Aristotelian position of pursuing its *Eudaimonia* (εὐδαιμονία) above and beyond all technical or craft-knowledge. On the other hand, technics is – for all its relative autonomy from an anthropomorphic regime of value – returned to its classical position as a mere *prosthesis* to the human. If Marx's communist future appears to envisage a form of human species-existence that exists entirely autonomously from technics, though, I would argue that this future is bought at the expense of a disavowal of what we have seen to be a central insight of his theory of labour, namely, the *mutual* constitution and evolution of the human and the technological across time.[31] Such is one fate of Marx's philosophy of technology: what was once an ontological condition – the co-constitution of the human and the technical in and through the labour process – is reduced to little more than a dialectical moment in the narrative of humanity's emancipation *from* technics.[32]

In seeking to read Marx 'beyond' Marx – if I may borrow a phrase from Negri – we must ask whether it is possible to purge his philosophy of technology of the teleology that seeks impossibly to liberate a collective *anthropos* from the constitutive *between* of the human/technical relation and begin to think freedom, justice, communism otherwise. It is, of course, already possible to detect some gestures in this direction both within and without contemporary Marxian thought (Derrida, Negri, Stiegler) although a question must remain about whether they go far enough. As Derrida argues, we cannot simply *oppose* technics and liberation: the virtualisation of time and space implied by the logic of *hauntologie* and made manifest in new media technologies obliges us to think what he gnomically calls another space for justice and democracy (*Specters of Marx*, p. 189). Yet, as Negri has argued, such a space remains at best a bare formal possibility in Derrida's own work on Marx, rather than the ground for a historical or material praxis (*Empire*, p. 217). For the authors of *Empire*, communism is famously not a utopic state that will arrive 'in a subsequent period' (*Marx Beyond Marx*, p. 181) but a possibility immanent to capital itself: what the real subsumption of the social creates is an immeasurable space and time, within capitalism, for a non-teleological communism of multiple, non-totalisable singularities (*Empire*, pp. 356–61). Not only does this Neo-Spinozan political ontology apparently force the recognition that 'there are no fixed or necessary boundaries' between 'the human and the machine' but also that 'nature itself is an artificial terrain open to ever new mutations, mixtures, and hybridizations' (p. 215). Such is why the political task of what Hardt and Negri call the multitude must consist in the actualisation of this potent machinic becoming within nature and the production of the form of the collective subject anew:

> ... it consists in an act, that is, of merging or hybridizing with the machines that the multitude has reappropriated and invented; it consists, therefore, in an exodus that is not only spatial but also mechanical in the sense that the subject is transformed into (and finds the cooperation that constitutes it multiplied in) the machine. This is a new form of exodus, an exodus toward (or with) the machine – a machinic exodus. (pp. 366–7)

If Hardt and Negri's immanent re-writing of the Fragment on Machines partially expunges Marx's own critique of its residual humanism – in which the subject is able to obtain emancipation from machinic labour – one might still argue that it only does so by re-instating that humanism at a higher level: what begins as a non-teleological affirmation of the multitude's machinic exodus seems to conclude with something that – for all its Deleuzean rhetoric of multiplicity – still looks very much like a collective human subject becoming 'an autonomous agent of production' with the capacity 'to take control of the processes of machinic metamorphoses' (pp. 405, 367). Just as Hardt and Negri risk re-instituting the Marxian opposition between the human and the machine in the very gesture of erasing it, so we will see in Chapter 6 that Stiegler, too, falls back into a classically Marxian critique of the contemporary industrialisation of time, memory and desire from the vantage point of (what at least *seems* to be) a more primordial anamnestic time: the human/technical complex is, once again, dialecticised, teleologised and anthropologised. Perhaps we might object here that, if we still wish to write, think and act in the spirit of Marx – in short to be 'Marxists' in any meaningful sense – we simply have no choice but to submit to such a teleology of emancipation: even Derrida, who could scarcely be called an orthodox Marxist, goes so far as to claim that 'if there is a spirit of Marxism which I will never be ready to renounce' it is 'a certain emancipatory and messianic framework' (*Specters of Marx*, p. 89). In Marx's philosophy of technology, though, it is possible to find another spirit of Marxism that might offer the following reply: Who – if anyone – is to be emancipated? What – if any – singular human essence or existence has been alienated or subjected? When – if ever – did that state of alienation commence? How – if at all – can it be overcome? And why – if there are no compelling answers to such questions – continue to speak of the emancipation of the human *from* technics at all?

3

What, to conclude, is Marx's legacy to the philosophy of originary technicity? It is difficult not to recall Derrida's words on the aporia of inheritance when faced with such an unanswerable question: '[a]n inheritance is never gathered together, it is never one with itself', he writes, 'one must sort out several different possibilities that inhabit the same injunction' (*Specters of Marx*, p. 16). As Derrida implies, and as this chapter has tried to show, Marx bequeaths more than one legacy to contemporary thinking about technicity: we are constantly compelled to reckon with (at least) *two* rival and competing voices within the philosopher's corpus that cannot be easily positioned on either side of an epistemological break, let alone sublated into some over-arching whole. To start with, Marx, *premier penseur de la technique*, is an indispensable precursor to Derrida's philosophy of originary technicity. By ontologising tool-use as the defining mode of human species-existence – even to the point of seeing the body itself as a kind of thermodynamic, energy-producing machine – he breaks decisively with the classical theory of technology: what-is-placed-in-front-of man, pros-thetically, for Aristotle, is placed *within* man, intra-thetically, for Marx. Just as Marxian philosophy of technology moves beyond hylomorphism, so the dynamic and metastable system of exchange it articulates between organ and environment also anticipates many twentieth- and twenty-first-century-theories about technics: Leroi-Gouhran's exteriorisation; Valera's structural coupling; Derrida and Stiegler's originary supplementation or prostheticisation. However, the problem is that Marx does not always follow the radical implications of this thesis for our concepts of human being, consciousness and philosophy all the way through: something – whether it be a residually humanist concept of a collective human essence of labour, a philosophy of alienation, a politics of emancipation – still precedes or exceeds technical constitution in his corpus from beginning to end. If this Aristotelian spirit never entirely goes away, though, I have tried to argue that it can only thrive by doing a violence to the radical *letter* of Marx's text: what Marx himself persists in describing as a fall *from* a collective human essence – into alienation, capitalisation, the machine – might more properly be described as a fall *within* that essence – which is thus revealed to be neither 'human' nor 'essential' in the first place. Perhaps, though, this aporia at the heart of Marxian philosophy of technology – man versus matter; idealism versus materialism; anthropocentrism versus technological determinism – is what really makes Marx our contemporary because, as we will see throughout this book, it is nothing less than the aporia of originary technicity itself. In the next chapter, though, we must turn to another important source for Derrida's philosophy of technics: Freudian and Lacanian psychoanalysis.

3
Psyche

> Everything fell into place, the cogs meshed, the thing really seemed to be a machine which in a moment would run of itself.[1]

In Sigmund Freud's vast corpus, the psyche is frequently said to operate like – perhaps even to *be* – a certain kind of machine. It is no exaggeration to say that 100 years after its inception, Freudian psychoanalysis still represents an indispensable reference point for continental philosophy and, in particular, theories of originary technicity. After Freud, we are no longer stable, independent and self-reflexive thinking subjects but the site of a complex of unconscious, autonomous, even inhuman drives and, in many ways, this insight ushers in the entire critique of the *ego cogito* in twentieth-century philosophy. For all its originality, though, psychoanalysis also pursues and extends a classically Cartesian thought experiment about the relation between the organic and the mechanical: what began as an analogy between the body and a clock in Descartes' *Meditations* reaches its logical conclusion in the Freudian comparison between the psyche and a machine.

To be sure, Freud first raises the possibility that unconscious processes might best be described in technological terms in the famous 1896 letter to Wilhelm Fleiss that forms the epigraph to this chapter: the psyche, he muses, resembles a machine that runs 'of itself' (*The Origins of Psychoanalysis*, p. 173). However this is only the beginning of what turns out to be a career-long exploration of the technical nature of the psychic apparatus. If we re-trace Freud's career from the materialist *Project for a Scientific Psychology* (1895), through the famous essay on 'The Uncanny' (1919) up to the suggestive thought experiment on the 'Mystic Writing Pad' (1927), we consistently find the father of psychoanalysis pursuing this analogy between the unconscious and technology up to – and sometimes beyond – its breaking point. In the course of 30 years of texts, psychic processes are variously compared to writing, signification and most intriguingly, the machine.

If Freud consistently raises the question of the originary technicity of the psyche, though, Jacques Derrida, Jacques Lacan and others have famously argued that he is unable or unwilling to push the analogy between mind and machine to its logical conclusion: Derrida famously argues that the unconscious might *materially* – and not just 'metaphorically' – be a species of writing. For Jacques Lacan – whose 1954–95 seminars announce themselves as a return to the materialism of early Freudian psychoanalysis – it is even possible to offer the very 'materialist definition of consciousness' that Freud ultimately refuses: consciousness is *indeed* a kind of machine – a ticking clock, a steam engine, an automatic camera, a cybernetic circuit or feedback loop – that operates, as it were, 'all by itself'.

This chapter explores psychoanalytic accounts of technology from Freud's *Project for a Scientific Psychology* through the essay on 'The Uncanny' to Lacan's landmark seminars on *Freud and Technique*. To what extent do Freudian and Lacanian psychoanalysis succeed in installing an originary technicity at the heart of the psyche – or, like Marxian historical materialism, do they also contain a residual pre-technological idealism?

1

In this opening section, I would like to briefly review Freud's life-long attempts to draw an analogy between psychic and technological processes or, more classically, between mind and machine. It is important to grasp at the outset, though, *why* Freud is compelled to articulate the structure of the psyche in technological terms at all. As he makes clear in the 1896 letter to Wilhelm Fleiss, Freud's entire project is premised upon the assumption that 'our psychical mechanism has come about by a process of stratification' (*Origins of Psychoanalysis*, p. 173). To put it another way, the psyche is not a monolithic entity but a quasi-Nietzschean field of force where perception and memory – the reception of new impressions and the retention of old ones by our nervous tissue – compete and interact with one another: 'the material present in the shape of memory-traces is from time to time subjected to a rearrangement in accordance with fresh circumstances' (p. 173). For Freud, what is absolutely new about this theory is the thesis that a memory is present not just once but *several times over* as it is progressively registered by different levels of the psyche: any impression must be recorded by (1) immediate perception, (2) the unconscious and (3) the preconscious before finally becoming (4) a conscious memory (p. 174). If memory is traced and re-traced at different levels – transcribed and re-transcribed – without being simply present in any one, then, it is wholly appropriate that Freud should begin to speak of the psyche in metaphorical terms as a kind of palimpsest where various forms of 'writing' – mnemic impressions – accumulate on top of each other without any one assuming priority: 'our psychic mechanism... is, as it were, transcribed' (p. 173). What form, though, does this psychic 'writing' take?

It is in the unfinished *Project for a Scientific Psychology* (1895) that we encounter Freud's first attempt to construct a theory of the psyche-as-writing machine. As is now well documented, Freud's biggest dilemma in building a theoretical model of the psyche is how to resolve the tension between two very different and apparently contradictory psychic functions: what enables our neurones to both *perceive* endless new impressions and permanently *retain* those impressions as memories at the same time? To explain this duality, Freud quickly departs from the physiological basis of the *Project* to offer a speculative hydraulic theory of the psyche based upon the relation between what he calls the ϕ neurones and the ψ neurones. On the one side, the permeable ϕ neurones are what enable us to perceive the external world. On the other side, the non-permeable ψ neurones are what allow us to retain traces of those perceptions. Yet it is the way in which our experiences are retained in memory that is most crucial to Freud's attempt to draw an analogy between the psyche and writing. Such a process of retention is described by Freud as a form of '*Bahnung*' – breaching, forging or pathbreaking – where intense experiences (such as pain) quite literally forge pathways within the ψ neurones. By arguing that experience leaves a literal trace on the psychic apparatus in the form of memory, Freud is able to draw a comparison between memory and writing: both effectively make a permanent mark upon a surface (*Origins of Psychoanalysis*, pp. 360–1). Now, as Derrida goes on to show in his classic essay 'Freud and the Scene of Writing',[2] we can arguably extend the analogy with writing further, by attending to the vital role of difference in Freud's model of the psyche. Perhaps most crucially from Derrida's perspective, what determines the whole process of memory retention in Freud's account is neither the qualitative nor even quite the quantitative substance *of* the traces but rather the 'ungraspable and invisible difference' *between* them ('Freud and the Scene of Writing', p. 201). For Derrida, to put it crudely, what is taking place here is a kind of psychic equivalent to the *arche*-writing he detects at work in Ferdinand de Saussure's theory of language: 'It is the difference between breaches which is the true origin of memory' (p. 201). Just as Saussure's theory of writing destroys any simple notion of representation, so Freud's differential model of the psyche also calls into question the very idea of a 'true origin' of memory that ever was or could be present: 'Trace as memory is not a pure breaching that might be reappropriated at any time as simple presence' (p. 201). If every memory trace only exists in relation to other traces, then there can be no original memory stored somewhere that corresponds to a certain event in our lives: 'we may still maintain that in the *first time* of the contact between *two* forces, repetition has begun' (p. 202). In this sense, Derrida argues, Freud opens up the possibility of a deeper – rather than merely analogical – technicity of the psyche: writing goes all the way down.

To Derrida's eyes, Freud's obsessive – indeed almost machinic (p. 198) – need to find technological *analoga* for the psyche continues throughout his career before reaching its conclusion in what is perhaps the best machine metaphor of all: a child's toy called the *Wunderblock* or 'Mystic Writing Pad'. It is here that the metaphorical gap between mind and machine begins to narrow to the point of non-existence (*Archive Fever*, p. 27). According to Derrida, as we will see later on, the reason why the machine can represent the psyche so precisely is because it is not a 'metaphor' at all but nothing other than a *material instantiation* of the psychic mechanism itself. First of all, though, we must return to Freud's own description of this ingenious writing machine:

> The Mystic Pad is a slab of dark brown resin or wax with a paper edging; over the slab is laid a thin transparent sheet, the top end of which is firmly secured to the slab while its bottom end rests on it without being fixed to it.[3]

Why is the Mystic Writing Pad – which enables the user to etch on the transparent cover sheet with a stylus, produce marks when the sheet is pressed against the wax below, and then erase those marks by lifting the sheet – the best trope for the psyche? For Freud, of course, the answer is that this child's toy perfectly reproduces the dual function that defines his theory of memory: it is capable of *both* the reception of new impressions on the top sheet *and* the retention of those impressions on the wax slab. If a sheet of paper preserves traces indefinitely but is quickly filled up, and a piece of slate is capable of receiving infinite traces but cannot preserve any of them, the *Wunderblock* is the only mnemo-technical apparatus capable of combining both infinite receptivity and permanent retention at the same time. Just as the Mystic Writing Pad enables one to write endlessly on the erasable top sheet whilst at the same time preserving traces of every inscription on the wax slab underneath, so the psyche is capable of receiving an infinite number of new impressions and permanently retaining the traces of those impressions in the unconscious: the *Wunderblock* is thus a 'concrete representation of the perceptual apparatus of the mind' ('Note on the "Mystic Writing Pad" ', p. 232). In Freud's conclusion, however, he goes on to make clear that – for all its heuristic value – this analogy between mind and machine still only holds *up to a point*:

> There must come a point at which the analogy between an auxiliary apparatus of this kind and the organ which is its prototype will cease to apply. It is true, too, that once the writing has been erased, the Mystic Pad cannot 'reproduce' it from within; it would be a mystic pad indeed if, like our memory, it could accomplish that. (p. 230)

For Derrida, in stark contrast to Freud, what we are dealing with here is less a mind/machine analogy – which must inevitably come to an end at some point – than something close to a *homology*: what we call 'mind' is itself a kind of machine at least in the sense that it is structured according to the autonomous and quasi-technical logic of repetition, supplementation and difference he famously calls *différance*. It is only necessary for us to trace the bare outline of this argument – which is in many ways a rehearsal of the critique of *anamnēsis* in 'Plato's Pharmacy' – before going on to focus on its larger implications. To begin with, Freud's entire argument is premised upon what Derrida sees as an untenable opposition between the living, spontaneous and infinite status of the psyche and its inert, dead technological supplement:

> In a letter to Fleiss, it will be recalled, Freud, evoking his representation of the psychical apparatus, had the impression of being faced with a machine which would soon run by itself. But what was to run by itself was the psyche and not its imitation or mechanical representation. For the latter does not live. ('Freud and the Scene of Writing', p. 227)

However, the psychoanalyst himself provides enough evidence to cast doubt upon this alleged spontaneity of the psyche: Freud describes in the very first sentence of the 'Note' how memory *necessarily* relies on a set of material aids to supplement its fallible powers ('A Note on the "Mystic Writing Pad" ', p. 227). Now, if memory simply does not possess the quasi-mystical powers of infinite, spontaneous reproduction that Freud attributes to it, then this means that it is already much closer to the allegedly inert, finite and non-spontaneous machine than he allows. If memory can no longer work 'all by itself', and what does not work all by itself is (at least in Freud's instrumentalist account of the *Wunderblock*) a tool, prosthesis or machine, then we can begin to conceive of our memory *itself* as a species of writing machine: '[f]ar from the machine being a pure absence of spontaneity, its *resemblance* to the psychic apparatus, its existence and its necessity bear witness to the finitude of the mnemic spontaneity which is thus supplemented' ('Freud and the Scene of Writing', p. 228). The supposedly extraneous technological supplement exposes a lack – an originary difference, spacing or deferral of presence – within the alleged state of nature that it supplements. This means that the machine is not so much an exteriorisation of the psyche, as Freud suggests, but what Derrida calls the sign of the existence of 'death and finitude *within* the psyche' (p. 228). In this account, memory is not just figuratively but quite literally a psychic *Wunderblock* – an originary writing or prosthesis stratified across a number of levels – that combines the preservation and deferral of traces, the living and the dead in the same machinic movement:

> A two-handed machine, a multiplicity of agencies or origins – is this not the original relation to the other and the original temporality of writing, its 'primary' complication: an originary spacing, deferring, and erasure of the simple origin, and polemics on the very threshold of what we persist in calling perception? (p. 226)

Why is Freud himself unable to make this radical move from seeing the machine as a metaphor for the psyche to seeing the psyche-as-machine? According to Derrida, of course, the psychoanalyst's final refusal of the mind/machine analogy is less an empirically verified conclusion than the logical outworking of an inherited set of Platonic oppositions about memory: *anamnēsis* and *hypnomēsis*; the true recollection and the false, technical substitute; the living memory and the dead, sophistical aide-memoire (*Meno*; *Gorgias*) (pp. 221–2). Despite his modernity, Freud scratches away on his wax block like Socrates drawing lines in the sand in the *Meno*, both already knowing what they will find there: the living, spontaneous immortality of *mneme*. To put it more bluntly, this little thought experiment turns out to be something of a self-fulfilling prophecy: Freud's search for material analogues for mind begins from the presumption that mind is not already material. By a rhetorical sleight of hand, the machine metaphor performs the exact opposite of its apparent purpose: it effectively de-materialises mind to the point where no exact material analogue for its organic, spontaneous and self-reproducing activity can ever be found. Yet, it is equally clear that this expanded concept of mind is mortgaged to an ingenious yet still reductive idea of the machine: Friedrich Kittler rightly observes that 'Freud's materialism reasoned only as far as the information systems of his era – no more, no less'.[4] For Freud, such a dogmatic affirmation of the original and essential immateriality of mind – such that no machine, however complex and sophisticated it may be, could ever properly reproduce it – effectively ensures that technology ends up being judged according to a set of criteria it can never meet. On the one hand, we see that all technological prostheses are to be measured exclusively in terms of whether they have the capacity to reproduce existing human functioning: auxiliary apparatuses like spectacles and ear trumpets are always constructed on the model of the organ – eye, ear – to be supplemented. On the other hand, we find that the immaterial functioning of memory can thus *never* be successfully reproduced in prosthetic form: mnemotechniques like wax, slate, paper and even the Mystic Writing Pad, as we have seen, only imperfectly materialise the mnemic organ ('Note on the "Mystic Writing Pad" ', p. 227). If the psychoanalyst can hardly be blamed for not being able to predict the revolutions in automated information processing of the next 75 years, though, the uncomfortable fact remains that this categorical rejection of the mind/machine analogy is something of a historical hostage to fortune based on nothing more secure than the limitations of the existing communication technology: the Mystic Writing Pad is 'no doubt infinitely more complex than slate or paper, less archaic than a palimpsest', as Derrida observes, but 'compared to other machines for storing archives, it is a child's toy' ('Freud and the Scene of Writing', p. 228).[5] Perhaps this means that Freud's 'Note on the "Mystic Writing Pad"' is not simply an allegory for the mind but for his own deeply instrumentalist understanding of all technologies (from ear trumpets to writing slates) that robs them of even the theoretical possibility of self-forming or autopoietic existence – within or without – the living psyche: what seems to 'run by itself' is never quite allowed to escape the grip of the thinking, remembering subject. In Freud's account, technological materiality seems to exist only in order to heuristically reveal the immateriality of mind.

In re-visiting Derrida's deconstruction of Freud's mind/machine analogy, we can see that his signal achievement is to take the machine from its position of notional exteriority and position it *within* the psyche as the non-essential 'essence' of memory. It is by no means a coincidence that Derrida himself goes on to observe that this gesture calls for a new 'question of *technology*' that exceeds the received oppositions between interiority and exteriority, psychical and non-psychical, life and death (p. 228): such claims undoubtedly represent the beginning of what, later in his career, will come to be known as originary technicity. Yet, it is also worth pausing briefly to flag up the question of whether something of Freud's anthropocentrism survives even in deconstruction; Derrida's philosophy certainly lays the groundwork for much of our critique of Freud – whether it be of the idealisation of psyche or of the reduction of technology to base matter – but perhaps we might be able to detect the *same* logic at work in deconstructive philosophy of technics. To make good his own position, for example, Derrida convicts Freud's 'Note' of harbouring a basic prejudice against worldly technology – and of the new archiving systems that far outstrip the capacity of the *Wunderblock* (p. 228) – but is deconstruction really much more interested in the *specificity* of this machine? For Derrida, in fact, what seems to be at issue here is not so much the concrete writing technologies themselves but rather the anterior state of difference, deferral or technicity before all empirical apparatuses that he calls '*arche*-writing' (*arche-écriture*):

> Writing, here, is *technē* as the relation between life and death, between presence and representation, between the two apparatuses. It opens up the question of technics: of the apparatus in general and of the analogy between the psychical apparatus and the nonpsychical apparatus. In this sense writing is the stage of history and the play of the world. (p. 228)

Such a vastly generalised state of originary technicity is, of course, always already in play as the condition of (im-)possibility of all historical or empirical technical apparatuses. Nevertheless, the fact remains that it is the Mystic Writing Pad – with its utterly singular conjunction of inscription and erasure – that gives Derrida (and Freud) access to this generalised state of *arche*-writing that is the play of the world. What opens the question of the technical apparatus in general may well be the originary technicity of the world but – and here arises the logic of the empirico-transcendental relation we raised in Chapter 1 – what makes possible the revelation of this generalised *technē*, in this context at least, is a specific technical apparatus. Just as Freud sees the machine as nothing more than a second-order materialisation of the immateriality of the psyche, however, there is a sense in which Derrida, too, tends to see it as merely a posterior materialisation, albeit this time of a prior, monolithic and generalised state of technicity: the historical writing pad itself effectively disappears into an immemorial logic of generalised writing. Both thinkers risk effacing what (to borrow a Heideggerian vocabulary) we might call the ontic state of matter that makes possible their larger ontological claim possible in the first place. Perhaps more revealingly still, we could argue that Derrida has every bit as instrumentalist an investment in the machine as Freud: what concerns both is solely what the *Wunderblock* has to tell us about the structure – natural or technical, living or machinic – of the psyche as opposed to anything within the technology itself. Either way, the empirical machine becomes little more than the object of a thought experiment, designed to disclose the prior status of mind, whether it be ideal or technical. If Derrida is right to take Freud to task for his naively anthropocentric account of the machine as an inferior prosthesis to the human, we might wonder whether there is still a kind of anthropocentrism in a reading that sees even the most sophisticated mnemic systems as nothing more than a liberated imitation of existing human capacities: 'we shall have to ask what the imitation, projected and liberated in a machine, of something like psychical writing might mean' (p. 199). In Chapter 5 of this book, we will see how Derrida's own attempt to 'materialise' mind is perhaps undermined by a corresponding tendency to 'idealise' materiality: what concerns him frequently still tends to be the *meaning* of matter for an absent but implied Cartesian subject – which is thus granted a curious kind of posterior life if only so that it might witness its own demise – as opposed to matter per se.

2

In his famous essay 'The Uncanny' (*das Unheimlich*) (1919),[6] Freud returns once again to the relation between the psyche and technology. It is arguably here, rather than in the note on the *Wunderblock*, that he comes closest to embracing the insight that many say is always latent in his work – the machinic status of the unconscious itself. As is now very well known, the German term '*das Unheimlich*' collapses together two apparently contradictory senses at once: 'the uncanny is that class of the frightening which leads back to what is known of old and long familiar' ('The Uncanny', p. 220). Why is it, though, that something can be *both* familiar *and* terrifying? To answer this question, Freud seeks to identify a specific case of the Uncanny and, of course, the one he alights upon could not be more pertinent:

> Jentsch has taken as a very good instance 'doubts whether an apparently animate being is really alive; or conversely, whether a lifeless object might not be in fact animate'; and he refers in this connection to the impression made by wax-work figures, ingeniously constructed dolls and automata. To these he adds the uncanny effect of epileptic fits, and of manifestations of insanity, because these excite in the spectator the impression of automatic, mechanical processes at work behind the ordinary appearance of mental activity. (p. 226)

For the German psychologist Ernst Jentsch, to whom Freud refers, we can witness this sense of the Uncanny at work in the nineteenth-century fantasy writer E. T. A Hoffman and, in particular, the latter's story 'The Sand-Man', in *Die Nachtstücke* (*The Night Pieces*). Nathaniel, the doomed protagonist of Hoffman's tale, falls deeply in love with a mechanical doll called Olympia constructed by his nemesis Coppola/Coppelius.[7] However, this automaton strikes one of Nathaniel's friends as deeply disquieting: 'We found Olympia to be rather uncanny [*Unheimlich*]. She seems to us to be playing the part of a human being.'[8] Finally, Nathaniel suffers a breakdown when he, too, realises that Olympia is indeed nothing more than a machine. If artificial life forms like Olympia – which fuse together nature and culture, the organic and the inorganic, the living and the dead in a deeply unsettling way – do indeed seem to fit the criteria of *das Unheimlich*, what interests Freud is precisely *why* they are uncanny. To what extent, in other words, does the sense of the Uncanny reside in the traumatised psyche of the observer – Nathaniel – or might it be generated by something within the technology itself?

It is hardly surprising by now to hear Freud's answer: 'I cannot think [that] the doll Olympia, who is to all intents and purposes a living being, is by any means the only, or indeed the most important, element that must be held responsible for the quite unparalleled atmosphere of uncanniness evoked by the story' (p. 227).[9] Again, Freud toys with the possibility of a relation between psychic and technological processes only to reject it. Why does the psychoanalyst reject out of hand the idea that the mechanical doll itself could be the locus of the uncanny? To start with, Freud observes that the figure of the doll is closely related to childhood life: children are especially prone to treating their dolls as if they were real. From this beginning in infantile desire, the psychoanalyst goes on to speculate that the mechanical doll's uncanniness is not the product of any immanent property within the doll itself but is rather an investment that has been placed in it by the human psyche. For Freud, as he goes on to show in a detailed re-reading of the Hoffman story, Nathaniel's sense of the uncanny upon the appearance of Olympia is in fact the product of a chain of psychic substitutions that leads all the way back to his childhood castration complex: Olympia the doll = Coppola, the optician who created her false eyes = Coppelius, the terrifying figure who used to visit Nathaniel's father late at night = the folk myth of the Sand Man who comes in the night to steal children's eyes = Nathaniel's father, who he fears will castrate him. Just as the Mystic Writing Pad is a materialisation of the psychic apparatus, in other words, so Olympia 'can be nothing else than a materialisation of Nathaniel's feminine attitude towards his father in his infancy' (p. 232n). Both lack any properly psychic capacity of their own because neither, in the final analysis, is really 'alive'. If we are still clearly in the realm of the mind/machine analogy here – which once again has the convenient effect of positing the essential immateriality of mind as a starting point – it is striking that the psychoanalyst again wants to push that analogy as far as it can go. In Freud's account, technology is seemingly never just one metaphor amongst others for the species of the Uncanny that is castration: the phenomena of unnatural or machinic motion – mechanical dolls walking and talking, dismembered feet dancing all by themselves – are depicted as almost definingly Uncanny (p. 244). Perhaps we might even go so far as to say that castration – the becoming-exterior of what is essential and the becoming-essential of what is exterior – has its logical conclusion in, and as, the machine.[10]

For Freud, though, the relation between the psyche and the machine becomes more intimate still when he turns to consider another species of the Uncanny, namely, the phenomenon of unconscious or voluntary repetition of the same activity:

> ...it is possible to recognise the dominance in the 'compulsion to repeat' proceeding from the instinctual impulses and probably inherent in the very nature of the instincts – a compulsion powerful enough to overrule the pleasure principle, lending to certain aspects of the mind their daemonic character... All these considerations prepare us for the discovery that whatever reminds of this inner 'compulsion to repeat' is perceived as uncanny. (p. 238)

To recall the original context, Freud is clearly alluding to the instinctive desire to return to the inorganic described in the roughly contemporaneous *Beyond the Pleasure Principle*,[11] but what is most intriguing here for us is the extreme proximity between the becoming-inorganic of the organic that is the death drive and the becoming-organic of the inorganic that is mechanical or automatic life. It is important to stress at this point that what Freud sees as uncanny is *not* the reactivation of some particular repressed content (such as dread of castration) but the bare phenomenon of repetition compulsion *itself* – what we might almost call the repetition of repetition (see Herz, pp. 101–2).[12] As we saw in the case of the Mystic Writing Pad, it is this originary repetition – what Derrida calls the originary trace that pre-exists and makes possible the very experience of the first time – that weds the psyche to the machine. Such an uncanny machinic dimension – which Freud himself always figures as a certain death[13] – at work within the living psyche can be witnessed in the phenomenon of the involuntary repetition of the same:

> ...it is only this factor of involuntary repetition which surrounds what would otherwise be innocent enough with an uncanny atmosphere, and forces upon us the idea of something fateful and inescapable when otherwise we should have spoken only of 'chance'. (p. 237)

Why does Freud identify repetition compulsion as uncanny but deny the same status to mechanical, automatic life when – by his own logic – both describe the same process of bare, subjectless repetition? By prematurely rejecting Jensch's concept of the Uncanny as an intellectual uncertainty concerning the animate/inanimate distinction, Freud fails to recognise how much it overlaps with his own definition of the Uncanny as repetition compulsion: Eric L. Santner makes this point very well when he says that '[t]here is nothing that throws more into question our status as living beings than the sheer, quasi-mechanical automaticity of the compulsion to repeat'.[14] If the point of chiasmus between mind and machine were not obvious enough already, Freud goes on to refer explicitly to the exact *same* phenomena cited by Jentsch earlier in his technological definition of the uncanny, namely, madness and epilepsy: 'The layman sees in them the working of forces hitherto unsuspected in his fellow-men, but at the same time he is dimly aware of them in remote corners of his own being' (p. 243). We saw earlier how Freud dismissed the possibility that automation – the becoming-live of the mechanical or the becoming-mechanical of life – constituted anything uncanny in itself: whatever uncanniness Olympia contained was the product of the castration complex rather than anything immanent to the technical apparatus. Now, though, this opposition between the human and the inhuman, psyche and matter, life and death has clearly become fraught: it may well be that the uncanny effect remains on the side of the living psyche, but the effect itself consists in precisely the unconscious recognition of the working of machinic forces *within* the psyche. In other words, Freud's Uncanny consists not just in 'doubts whether an apparently animate being is really alive', nor over 'whether a lifeless object might not be in fact animate', but in what Lacan describes as the more disturbing and hyperbolic doubt that I, too, might be a machine:

> We are very well aware that the machine doesn't think. We made the machine, and it thinks what it has been told to think. But if the machine doesn't think, it is obvious that we don't think either when we are performing an operation. We follow the very same procedures as the machine. (*Seminar II*, p. 304)

In Freud's account, of course, the Uncanny in all its myriad and contradictory forms is ultimately a symptom of the 'return of the repressed': what makes the Uncanny both familiar and terrifying, utterly alien yet somehow always already known, is the becoming-conscious of unconscious desire ('The Uncanny', p. 245). It is crucial to ask, though, what *exactly* is being repressed, and returning, in the phenomenon of repetition compulsion. As we have already begun to see, the answer is emphatically not some particular psychic memory – an infantile desire or fear for example – but quite simply the blind, empty, ongoing phenomenon of repetition itself: 'whatever reminds of this inner "compulsion to repeat" is perceived as uncanny', Freud clearly writes. Another way of putting this is to say that what is going on (and on) in repetition compulsion does not have its origin in some repressed traumatic event but the other way around: a singular event is what reminds us of our *compulsion to repeat.* We are reminded once more that the psyche is not primarily a 'thing' – something composed of memory traces that were, could have been or will become present or re-presented – so much as a subjectless *technē* of repetition itself. Yet, what remind us of these mechanical processes are, appropriately enough, real, concrete machinic phenomena: automata, living dolls, writing apparatuses, people experiencing involuntary fits and so on. What, then, exactly is the function of machines in Freud's account of the Uncanny – metaphors for mind, materialisations of infantile dread or something else? To put it somewhat tautologically, I wish to suggest that what is supposedly being 'materialised' in the figure of the mechanical doll, the epileptic or the madman is nothing other than the materiality of the unconscious itself:

> ... these [mechanical phenomena] excite in the spectator the impression of automatic, mechanical processes at work behind the ordinary appearance of mental activity. (p. 226)

Perhaps the only thing that stops us from grasping the materiality of the unconscious itself is once again the basic metaphysical anthropocentrism that drives Freud's thought experiment. Just as the material 'metaphor' that is the Mystic Writing Pad exists in order to retrospectively prove the 'literal' immateriality of memory, so the mechanical analogue Olympia becomes a kind of psychic scapegoat for Nathaniel's own machinism: the imitation produces the real. If machines are metaphors, and Olympia is a machine, then Nathaniel must logically be posited as the literal organic meaning of that metaphor – even when Hoffman's hero dreams of the Sand Man unscrewing his own arms and legs as if he, too, were nothing but a living doll (p. 232). For Freud, in other words, what is taking place here is a classic gesture of technological abjection: technology is devolved from its anterior position as the structural condition of the psyche to nothing more than a posterior material analogue for a – de facto de-materialised – unconscious.

What exactly, to conclude, is taking place in Freud's Uncanny? It is now almost something of a cliché to state that Freud's essay is *itself* an experience of the Uncanny: everything he says about undecidability, the double, metaphor and so on is also embodied by the strange, circumlocutory itinerary of his own argument. According to the same logic, we can detect a certain uncanny repetition compulsion or self-reflexiveness, in the psychoanalyst's constant return to figures of the machine itself – Olympia, Nathaniel, the madman, the epileptic. Nevertheless, at the same time, the psychoanalyst is always attempting to short-circuit this feedback loop by insisting upon the primary, organic status of the psyche itself. For Freud, as we have seen on a number of occasions throughout this chapter, it is essential to preserve a minimal difference between mind and machine, even or especially as that opposition collapses all around him, and this difference becomes the site of 'the human' itself. Now, we can put to one side his constant deployment of the machine as a pure metaphor here – which obviously presumes the existence of some notional non-machinic 'reality' – because, even when the metaphor is exploded, the difference remains. Just as the 'Note on the Mystic Writing Pad' always insists upon the supposed spontaneity of the psychic apparatus – in contrast to the inert materiality of the writing machine – so 'The Uncanny' seizes upon a very familiar ghost in the machine of consciousness to uphold the opposition between the human and the automaton even in the act of laying it bare: '*whatever reminds of* this inner "compulsion to repeat" *is perceived as* uncanny' (p. 238, emphases mine). If the human is a machine, in other words, it is a machine that *knows* it is a machine – that is capable of consciously reflecting upon its own materiality – and this minimal residue of a transparent self-cognisant Cartesian *ego cogito* is enough to maintain its exceptional status. Perhaps more importantly still, Freud's residual idealism here also ends up – in a strangely similar way to Marx in Chapter 2 – returning the machine to its traditional position as quasi-Aristotelian supplement to, or instrument, of the human: technics, whether we see it as the reification of labour power (Marx) or an analogue for the psyche (Freud), remains nothing but a material expression of pre-existing energies – even when those energies are themselves revealed as material. In this way, Freud's mind/machine tautology still manages to secure a certain attenuated priority for mind *over* the machine: technics is never anything more than a secondary materialisation of mind's own materiality and exists purely in order to remind us of the (higher) species of machines that we ourselves are.

3

Finally, and more briefly, I would like to turn to another psychoanalytic attempt to posit the materiality of mind: Jacques Lacan's so-called 'materialist definition of consciousness' in his *Seminar II: The Ego in Freud's Theory and in the Technique of Psychoanalysis 1954–5.* To put it in Lacan's own words, what this landmark early seminar seeks to address is nothing other than the classic eighteenth-century question of the relation between life and the machine: 'why are we led to think of life in terms of mechanism?' (*Seminar II,* p. 31). It is crucial, of course, to set this specific question against the backdrop of the psychoanalyst's broader intellectual project at this stage of his career, namely, the critique of the Cartesian philosophy of consciousness that survives all the way up to phenomenology (Husserl, Merleau-Ponty) and Ego Psychology (Hartmann).[15] As Lacan makes clear throughout *Seminars I* and *II,* the crucial insight of Freudian psychoanalysis from the *Project for a Scientific Psychology* (1895) up to *Beyond the Pleasure Principle* (1918) is the existence of the *unconscious* as the engine of subjectification: what we call the 'I' is not the transparent and self-cognisant 'I think' or *ego cogito* of Cartesianism but something that is split or bifurcated by desire. For Lacan, in other words, what the early Freud teaches us is that it is necessary to give a new description of the phenomenon of consciousness as something that exists quite independently of any experience of the thinking ego: 'the issue is to how to free our notion of consciousness of any mortgage as regards the subject's apprehension of itself' (*Seminar II*, p. 57). If we are to articulate this seemingly paradoxical concept of a 'consciousness without ego' – where the 'I' emerges from essentially opaque, unconscious desires – Lacan argues that we must return to Freud's persistent attempts to draw an analogy between the psychic apparatus and technical or mechanical devices: 'He [Freud] realises that the brain is a dream machine' (p. 76). In Lacan's view, the figure of the machine – the automaton that functions all by itself – instantiates the phenomenon of a consciousness that operates without requiring the constituting unity of an ego. Such, at least, is what Manfred Frank caustically calls the defining 'dream' of structuralism: 'a machine without a subject'.[16]

First of all, Lacan prepares the ground for his self-styled materialist definition of consciousness by sketching a brief history of the mind/machine analogy up to Freud's own time. What, he asks, led Descartes to inaugurate the era of modern biology by thinking of the human body as a kind of mechanism in the *Treatise on Man*?[17] As Lacan shows, the deceptively simple answer to this question was the invention of a very specific mechanism: the clock. Not only was the clock a machine that 'worked by itself', but it was something 'which could embody in a quite striking way something essentially human' (p. 73). Quite simply, clock-time, which is something utterly essential to human Being-in-the-World – we all need to know the time (p. 74) – but which nonetheless operates wholly independently *of* humans, embodies the extent to which our subjectivity is constructed out of external *symbols*: 'man is always cultivating a great many more signs than he thinks' (p. 122). To Lacan's eyes – and here we can begin to detect the trace of his famous linguistic re-writing of Freud – what the clock and later machines make possible is the recognition of the extent to which the human psyche is already committed to – devolved onto – an autonomous symbolic order:

> What is at play in the machine?... It isn't a simple artefact, as could be said of chairs, tables, and of other more or less symbolic objects, among which we live without realising that they make for our own

> portrait. Machines are something else. They go much further in the direction of what we are in reality . . . The machine embodies the most radical symbolic activity of man. (p. 74)

Secondly, Lacan moves on from Descartes to draw an intriguing historical contrast between Hegel and Freud's concepts of subjectivity. On the one extreme, he argues that Hegelian phenomenology represents the furthest extension of a theory of consciousness as pure knowledge: man is nothing but the accumulation of past knowledge. On the other extreme, he suggests Freudian psychoanalysis proposes a theory of consciousness as non-knowledge: man becomes a kind of unknowing machine. If Hegel and Freud represent two opposing poles of subjectivity – the one knowing, the other non-knowing, the one conscious and the other unconscious, the one accumulative and the other mechanically repetitive – then the obvious question is what changes in the 100 years between them and Lacan's answer is remarkable: 'there's the advent of the world of the machine' (p. 179). Just as the clock revealed to Descartes that human subjectivity is built out of symbols – we are, as it were, the effects of our own outside – so industrial machines like the steam engine and, in particular, the thermodynamic calculus of input/output it reveals stand behind the Freudian theory of subjectivity as the site of a series of mechanical, self-regulating, homeostatic drives like the Pleasure Principle:

> Energy, I had you observe last time, is a notion which can only arrive once there are machines . . . There are no examples of energy calculations in the use of slaves . . . It took machines for us to realise they had to be fed. And more – they had to be looked after. But why? Because they tend to wear out. Slaves do as well, but one doesn't think about it, one thinks that it is natural for them to get old and croak. And later on, it dawned on people, something which was never thought of before, that living things look after themselves all on their own, in other words, they represent homeostats. (p. 75)

Such an attempt to think the psyche in terms of the machine is once again an attempt to cleave apart consciousness from any self-cognisant ego: the psyche-as-machine can run perfectly well – ticking like a clock, pumping like a steam engine – without any self-reflexive or egoic experience whatsoever. Perhaps most precisely, what the machine reveals is that the psyche is always already given over to a circuit of symbols working, so to speak, all by itself: what appears to be *outside* the psyche – language, the sign, the technical or material per se – is once again located *within* it as its essential structure. For Lacan, it is this machine that enables the Freud of *The Interpretation of Dreams* to break with the still residually vitalist or organic models of energy that, he argues, still dominate the 1895 *Project* and to recognise that language goes all the way down:

> He realises that the brain is a dream machine. And it is in the dream machine that he rediscovers what was there all along and which hadn't been noticed, namely, that it is at the most organic and most simple, most immediate and least manageable level, at the most unconscious level, that sense and speech are revealed and blossom forth in their entirety. (p. 76)

In this immediate relation between the organic and the inorganic, we can begin to glimpse the early Lacan's fundamental insight into the relation between the psyche and signification: the unconscious is structured like a language, which is to say, a circuit of signifiers, which is to say, a machine.[18]

To Lacan's way of thinking, the autonomy of consciousness from any experience of the ego is articulated most vividly for the participants of his seminar in the form of a (surprisingly still somewhat obscure) thought experiment. Let me recite this fascinating little heuristic fiction in full:

> What is the image in the mirror? The rays which return on to the mirror make us locate in an imaginary space the object which moreover is somewhere in reality. The real object isn't the object that you see in the mirror. So here there's a phenomenon of consciousness as such. That at any rate is what I would like you to accept, so that I can tell you a little apologue to aid your reflection.
>
> Suppose all men have disappeared from the world. I say *men* on account of the high value which you attribute to consciousness. That is already enough to raise the question – *What is left in the mirror*? But let us take it to the point of supposing that all living beings have disappeared. There are only waterfalls and springs left – thunder and lightning too. The image in the mirror – the image in the lake – do they still exist?
>
> It is quite obvious that they still exist. For one very simple reason – at the high point of civilisation we have attained, which far surpasses our illusions about consciousness, we have manufactured

> instruments which, without in any way being audacious, we can imagine to be sufficiently complicated to develop films themselves, put them away into little boxes, and store them in the fridge. Despite all living beings having disappeared, the camera can nonetheless record the image of the mountain in the lake, or that of the Café de Flore crumbling away in total solitude. (p. 46)

What is the meaning of this strange mechanical tableau? It is in this startling image of an automatic camera taking photographs in a post-human world that we find the materialist definition of consciousness par excellence. According to Lacan, what is taking place when the camera automatically records the image of the mountain in the lake is nothing less than a *phenomenon of consciousness*: the camera is conscious *of* the image. Despite its materiality, the camera is capable of recording the image of the mountain in its viewfinder in the same way that the human eye can perceives an image of a real object in the mirror. Both the camera and the eye have the singular ability to capture the real, the imaginary and the symbolic all at once. However, it is only possible to attribute something like 'consciousness' to the camera here if we transform our understanding of consciousness itself and, of course, this is the whole point of the thought experiment. If there is still something radically counter-intuitive about conceiving of a camera as capable of thought, Lacan argues, it is only because we habitually equate consciousness with *self*-consciousness, that is to say, with the unifying form of the ego as the agent of that phenomenon. Now, in stark contrast to this view, Lacan seeks to pursue Freud's radical insight in 'The Ego and the Id' that consciousness is little more than a kind of lens, screen or surface onto which the unconscious projects itself: this is why he is able to make the otherwise shocking claim that consciousness can occur on any surface – a lake, a mirror, a camera lens – that is capable of reflecting an image (p. 49).[19] 'So-called Man, distinguished by his so-called consciousness, is unnecessary for this process', glosses Friedrich Kittler, 'because nature's mirrors can accommodate these types of representation just as well as the visual centre in the occipital lobe of the brain' ('The World of the Symbolic', p. 131). For Lacan, in other words, what is happening in the story of the camera is nothing less than the liberation of a material, utterly ego-free and subjectless *technē* of consciousness:

> This is what I want you to consider as being essentially a phenomenon of consciousness, which won't have been perceived by any

> ego, which won't have been reflected upon in any ego-like experience – any kind of ego and of consciousness of ego being absent at the time.
>
> You'll tell me – *just a minute though! The ego is somewhere, it's in the camera.* No, there's not a shadow of ego in the camera. (p. 47)

Perhaps most importantly of all, Lacan concludes this thought experiment by characteristically turning the inhuman gaze of the camera back upon human consciousness itself: what concerns him is not that machines can be as conscious as humans, but that human consciousness is *itself* a kind of machine that operates independently of any experience of the ego. This is the real meaning of the materialist definition of consciousness. We humans have not really 'disappeared' after all – for the camera is *us*, human consciousness is itself a kind of automatic photography. In Lacan's theory, the automatic camera – endlessly taking pictures without ever being aware of what it is doing – becomes the latest exemplum of a theory of the psyche whereby human subjectivity is the effect of the symbolic order:

> I am explaining to you that it is in as much as he is committed to a play of symbols, to a symbolic world, that man is a decentred subject. Well, it is with this same play, this same world, that the machine is built. The most complicated machines are made only with words [*ne sont faites qu'avec des paroles*].
>
> Speech is first and foremost that object of exchange whereby we are recognised, and because you have said the password, we don't break each other's necks, etc. This is how the circulation of speech begins, and it swells to the point of constituting the world of the symbol which makes algebraic calculations possible. The machine is the structure detached from the activity of the subject. The symbolic world is the world of machine. (p. 47)

To what extent, though, does Lacan's materialist definition of consciousness manage to resist the residual humanism we found in Freud or – like Derrida – does he remain complicit with what he attempts to critique? It will already be clear even from my own very brief account that Lacan deeply radicalises the relation between mind and machine. As his experiment shows, what we are dealing with here is no longer an analogy between psychic and technical apparatuses – where the camera is merely a rhetorical figure or trope for the actual condition of consciousness – but another kind of homology. Both mind *and* machine are material. To put it in Lacan's own brutally clear words, in response to a question from Octave Mannoni, the human and the machine are both the effect of a radically inhuman, cybernetic other that operates independently of any agency:

> Don't be soft. Don't go and say that the machine is really rather nasty and that it clutters up our lives. That is not what is at stake. The machine is simply the succession of little 0s and 1s, so that the question of whether it is human or not is entirely settled – it isn't. Except, there's also the question of knowing whether the human, in the sense in which you understand it, is as human as all that... the notion of humanism... seems to me to be for us sufficiently weighted down by history so as to be capable of being considered as a particular position within a very much localised field which we imprudently continue to call humanity. And we shouldn't be surprised by the fact that the symbolic order is absolutely irreducible to what is commonly called human experience. (pp. 319–20)

Yet, if consciousness-as-machine seems to liberate the subject from any residue of Cartesianism, it is worth pausing to consider whether the ghost in this machine disappears quite as easily as we might think. First of all, Lacan *himself* is permanently on the alert against the spectre of what he calls psychoanalytic 'entification' (p. 12): the temptation to presume the existence of a *substance* – the soul, the *ego cogito* or some other such homunculus – lying behind the psychic field of force (pp. 11–12). Such is the *caveat emptor* that Lacan issues not only to Ego Psychologist Heinz Hartmann – who notoriously posits the possibility of an autonomous ego that can somehow transcend all conflict (pp. 11–12) – but to any participant in his own seminar (in particular Leclaire (pp. 53, 55, 62), Lefebvre-Pontalis (p. 58) and Mannoni (p. 319)) who is tempted to essentialise what is nothing more than an *imaginary* subjective unity:

> The subject is no-one. It is decomposed, in pieces. And it is jammed, sucked in by the image, the deceiving and the realised image, of the other, or equally by its own specular image. That is where it finds its unity. (p. 54)

By retrospectively attributing the specular ego-effect produced by consciousness-as-machine to that machine as its motive cause, however, Lacan's own theory of the psyche risks committing the very Cartesian 'idolatry' (p. 53) he is warning against. For Serge Leclaire, one of the participants in the 1954/95 seminar, Lacan's refusal to entify the subject only ever leads him to erect some *other* entity into the position of the *ego cogito* and thus we end up right back where we started:

> I have the impression that in refusing this deliberate entification, the subject, we have a tendency, and you have a tendency, to carry this idolisation over to another point. At this moment, it won't be the subject, it will be the other, the image, the mirror. (p. 55)

If Lacan seeks to materialise the ego out of all existence through his parable of the camera, in other words, Leclaire argues that he ends up ego-ising – idealising – matter itself in what amounts to a new Cartesian syllogism: 'I photograph therefore I am a materialist consciousness.' Now, we already know, of course, that Lacan would not accept this critique for a moment: the whole point of the camera, as we have seen, is that there is not a shadow of ego within this pure mechanism of reflexivity. What does seem to be true, though, is that there *is* the shadow of an ego apprehending the camera somewhere in this little drama, as another distinguished participant in the seminar, Jean-Bertrand Lefebvre-Pontalis, observes. Perhaps the most important – and neglected – aspect of the whole story of the camera, Lefebvre-Pontalis argues, is that human beings must *come back into the world* in order to apprehend the materiality of their own consciousness:

> I never said that consciousness entirely exhausted subjectivity, which would in any event be rather difficult given phenomenology and psychoanalysis, but simply that the cogito represented a sort of model for subjectivity, that is to say, rendered quite apparent the idea that there must be somebody for whom the word like has a meaning. And that you appeared to omit. For when you went into your little apologue about the disappearance of men, you forgot one thing, which is that men had to return in order to grasp the relation between the reflection and the thing reflected. Otherwise, if one considers the object in itself and the film recorded by the camera, it is nothing other than an object. It isn't a witness, it's nothing... I really can't see why it is so important to demolish consciousness if it is to bring it back in at the end. (p. 58)

Why, in Pontalis' view, is Lacan compelled to bring back humanity – like a kind of *Deus ex machina* lowered onto the stage at the conclusion of a classical drama to restore mortal order – to judge that the camera is indeed conscious? Just as Freud brings back the ego when he defines the Uncanny as the *awareness* that consciousness is machinic, so Lacan, in turn, is forced to re-introduce some quasi-egoised phenomenon that is able to reflect upon consciousness-as-machine. On the one hand, the psychoanalyst expunges all traces of human life from the planet. On the other, though, he must resurrect humanity in order to demonstrate to it the possibility of machine consciousness (*Was ist Neostrukturalismus?*, p. 398).[20] The ego must return in order for us – Lacan, his students, his readers – to become conscious *of* the material basis of consciousness. There must be a super-added phenomenon for which the photographic image has a meaning – which is able to apprehend its status *as* a witness to a real object – and 'consciousness' is as good a name for that phenomenon as any other. This is why Lacan's materialist definition of consciousness might not quite be material after all: its machinic, unreflexive materiality must still be adjudged at the tribunal of an absent, but implied, *ego cogito* for whom the word 'like' has a meaning. In the end, consciousness only 'returns' because, in a more important sense, it has been waiting in the wings all along: it is required even if only to apprehend its own materialisation or autonomisation as such.[21]

In a larger sense, then, I want to suggest that – for all its originality – Lacan's materialist definition of consciousness remains a story told by humans, about humans, for humans: the whole drama has been staged in order to reveal our own symbolic constitution to ourselves. It is probably inevitable that such an explicit critique of Cartesianism has to be fought on 'enemy territory' – which is to say the absolute centrality of consciousness to any definition of subjectivity has to be its starting point – but the risk is that, like Freud before him and Derrida afterwards, Lacan remains too close to what he is criticising: the phenomenon of consciousness – whether ideal or material, living or machinic – is centre-stage at all times. To be sure, consciousness is exteriorised, materialised and technologised – our identity is displaced onto an inhuman symbolic outside – but there is always a lingering sense in which that outside is deemed to exist simply as a means *towards* a new act of subject constitution:

> Why are we so astonished by these machines? It may have something to do with the difficulties Freud encountered. Because cybernetics also stem from a reaction of astonishment at rediscovering that this human language works almost by itself, seemingly to outwit us. (p. 119)

Such grand gestures of machinic triumphalism cannot wholly obscure the fact that technological exteriority is still being subtly instrumentalised – internalised – to the point where its sole purpose is to make manifest the prior, linguistic, displacement of the subject. Yet, as always, there is another, obscured, side to this question: the radically non-subjective status of the machine itself. For Lacan, just as for Freud and Derrida, the unremitting focus on what the machine means *for* a certain idea of the subject comes at the cost of a foreclosure of any technological becoming that could occur independently of that subject: the possibility of technological autonomy or determinism never arises, not simply because he could have had no knowledge of it at the historical time of writing,[22] but because in the final analysis he has no interest in it. Just as a certain Marx cannot see the machine as anything other than the alienation of human labour power, and a certain Freud cannot see it as anything but the technological reproduction of memory, so a certain Lacan sees it as nothing but the materialisation of the psyche or, better, the material confirmation that the psyche is *already* material: technological exteriority remains nothing more than the anthropomorphised projection of psychic interiority. Perhaps this is why Lacan's philosophy of technology must – for all its interest in machinic consciousness – remain committed to a basic Aristotelian instrumentalism: what is interesting about the machine from his perspective is always the fact *that humans made it* and thus something essentially 'human' endures within it. If Lacan has a theory of originary technicity at all, it is entirely modelled upon his prior theory of consciousness as the projection of unconscious agency: whatever serves to describe the phenomenon of consciousness – an egoless, unreflexive materiality – must also serve to describe the phenomenon of technology no matter how reductive such an analogy may become on both sides. The psychoanalyst must deprive the machine of any historical potential to evolve beyond the most limited, unreflexive phenomenon of consciousness in order to make good his materialist reduction of the psyche. This leads us to the uncomfortable conclusion that the early Lacan's attempt to use the machine to 'free our notion of consciousness of any mortgage as regards the subject's apprehension of itself', as he puts it, is *itself* mortgaged to a historically finite concept of technology as non-reflexive, and one that was already in the process of becoming (at least theoretically) obsolete. All this means that – for all his fascinating thought experiments – we leave Lacan with a lingering sense of a missed opportunity. What, if I may indulge in a little counter-factual speculation, might have happened if the psychoanalyst had tried to construct a thought experiment that took the becoming-mind of matter as seriously as it did the becoming-matter of mind? What if – rather than a clock, a steam engine or an automatic camera – Lacan had taken, say, Benjamin's cinema or, better still, Turing's hypothetical thinking, intelligent computer as his material model for the psyche? What would an exchange between Lacan's consciousness-as-machine and Turing's machine-as-consciousness – the human camera interacting with the robotic human – look like?[23]

4

What, to draw this chapter to a close, are we to make of Freud and Lacan's machines? Firstly, and most importantly, Freud remains – for all his hyperbolic speculations, unscientific methodologies and technological limitations – an essential thinker of what Derrida calls the originary technicity of consciousness: what we today call the discipline of artificial intelligence – with all its attendant debates about consciousness, self-reflexiveness, embodiment, distribution and affect – finds much of its space already mapped out 100 years ago in psychoanalysis. It is even possible to find something of this technicity in Freud's constant return to figures of the machine – writing, the Mystic Writing Pad, the automaton, even the figure of the psychoanalyst himself – to the point where they exceed their own figural status. As we have seen throughout this chapter, psychoanalysis does not just observe the technical constitution of perception, memory and so forth but *perform* it: the recursivity of the machine metaphor is *itself* machinic (see 'Freud and the Scene of Writing', p. 138). Everything falls into place, the cogs mesh, the thing really seems to be a machine running all by itself. Yet, this is not quite the whole story either, because there is another kind of machine working all by itself throughout the Freudian text too – what Agamben calls the 'anthropological machine' that, just as automatically, recursively and unthinkingly, seeks to assert the exceptionality of the human at all costs (*The Open*, pp. 33–8). We saw in the previous chapter how historical materialism insists upon the essentially human character of labour – we are apparently the only being who labours even if, in a deeper sense, we are ourselves the products of labour – and now we can see that psychoanalysis assigns an equally problematic defining role to thought. For Freud, the last redoubt of the human becomes a kind of spectral *ego cogito*: the thinking subject is granted an impossible afterlife in which it is able to bear witness to, apprehend and be conscious of its own material obsolescence. If Freudian psychoanalysis thus never entirely dispels the Cartesian ghost in the machine – the ghost *as* machine – we can also detect something of the same idealist recidivism in succeeding theories of the psyche: 'I don't much like hearing that we have *gone beyond*... Descartes', Lacan presciently notes, 'We go beyond everything and always end up in the same place' (*Seminar II*, p. 71). Just as Freud defines the psyche as the *awareness* of its own mechanical repetition as such, so Lacan invokes the *consciousness* of materiality as much as the *materiality* of consciousness. By recalling and adapting Agamben's claim that 'man has no specific identity other than the *ability* to recognise himself' (*The Open*, p. 26), we might go so far as to say that, for Lacan, we are the only beings who *know* there is no difference between us and machines, and thus our minimal idealist difference *from* the machine is preserved even in the moment of its erasure. Perhaps, as we will see in more detail later on in this book, even Derrida's deconstruction of the mind/machine difference preserves this idealist residue: what is never entirely questioned is that this deconstruction remains a condition whose implications can at least partly be recognised, reflected upon and *thought* by a problematised but still identifiable *ego cogito*. In the following chapter, however, we turn to another key precursor for Derrida's philosophy of originary technicity that seeks to expunge what it sees as the residual metaphysical humanism of both historical materialism and psychoanalysis alike: Martin Heidegger's ontology.

4
Being

> Technology [*der Technik*] is not equivalent to the essence [*Wesen*] of technology.[1]

In the view of both devotees and critics alike, Martin Heidegger is the single most influential twentieth-century philosopher of technology.[2] It is Heidegger (even more powerfully than, say, Marx) who is deemed to preside over the final transformation of technology into ontology. As Richard Rojcewicz has recently argued, what Heidegger shows us is that the philosophy of technology is 'actually *equivalent* to first philosophy, since, for him, technology is nothing other than knowledge of what it means to be in general'.[3] To put it in the philosopher's own dramatic words from the *Basic Questions of Philosophy*, what the Ancient Greeks called '*technē*' ultimately meant neither a purely technical skill or 'know-how', nor a kind of instrument or artefact, but quite simply the ability 'to grasp beings as emerging out of themselves in the way they show themselves, in their essence, *ειδος* [*eidos*]'.[4] For Heidegger, the history of the philosophy of technology from Aristotle to the epoch of contemporary techno-science effectively becomes the history of Being's *own* self-disclosure – a disclosure that changes radically over time – to that being who is most equipped to receive it: *Dasein*. Such a history begins, as we will see, with the ancient idea of technology as *poiēsis* (*Hervorbringen* or bringing-forth) and comes to an apparent conclusion in the predominant modern theory of technology as what Heidegger famously calls *das Gestell* (Enframing).

If Heidegger is undoubtedly another essential interlocutor for Derrida's own philosophy of originary technicity – insofar as he raises technics to a level where it is coterminous with the question of Being itself – it is striking that, like Marx and Freud before him, his ontology has consistently been criticised for performing a certain reduction or exclusion of technics. On the one hand, his re-reading of the history of philosophy is charged with valorising a quasi-naturalist or even ecological concept of *phusis* (together with its distinctively poietic mode of presencing) over and against the autonomy of *technē*. On the other hand, his existential analytic is accused of perpetuating the reign of a quasi-idealist thinking subject for whom technology is subordinate to a pre-technological ontological questioning. Perhaps the most important corollary of this exclusion, it is deemed, is a repression of what Heidegger calls 'ontic' or empirical technics as a constitutive force within any ontology: ontic technology is, apparently, permitted no meaningful role in its own ontological questioning.

This chapter re-visits Heidegger's questioning concerning technology from his *destruktion* of the original Greek concept of *technē*, through the analytic of *das Zeug* (equipmentality) in *Being and Time*, all the way up to the famous critique of *das Gestell* in the essays of the 1940s and 1950s. To what extent does Heidegger's focus on what he calls the 'essence' (*Wesen*) of technology over and above technics itself give us a means of *thinking* technology for what is almost the first time or, as Derrida famously objects, does it preserve 'the possibility of a thought that questions... from any original and essential contamination by technology'?[5]

1

In many ways, Heidegger's radical re-reading of the original Greek concept of *technē* can only come as a shock to any reader schooled in what has become known as the 'philosophy of technology': everything we thought we knew about technology – it is an inert, neutral prosthesis or artefact, possessing no natural capacity for autonomous motion, utilised by an efficient cause – turns out to be wrong. To understand why, Heidegger's reading of *technē* has to be placed against the backdrop of his general re-evaluation of the Pre-Socratic concept of *phusis*. For Heidegger, according to one of his most famous discussions of the term in the 1935 lectures published as the *Introduction to Metaphysics*, what the Ancient Greeks call *phusis* names not some ontic being or region such as 'nature', but nothing less than the 'emerging-abiding' sway of Being itself:

> Now what does the word *phusis* say? It says what emerges from itself (for example, the emergence, the blossoming, of a rose), the

> unfolding that opens itself up, the coming-into-appearance in such unfolding, and holding itself and persisting in appearance – in short, the emerging-abiding sway.[6]

Such an ontological concept of *phusis* as Being is prior to any particular being or process that may take place like the blossoming of a flower or the birth of a baby because it bespeaks the originary donation of the 'beingness' of those beings: '*Phusis* is Being itself, by virtue of which beings first become and remain observable' (*Introduction to Metaphysics*, p. 15). If the Ancient Greek concept of *phusis* has been increasingly reduced to the Latin 'nature' (*natura*); the Christian 'creation' (*ens creatum*) and even the modern scientific concept of a material organism, Heidegger insists on its irreducibility to its own reception history. Far from being a simple manifestation of nature, *phusis* is the basis on which the natural world is disclosed to us as such: 'what they had to call *phusis* disclosed itself to them', writes Heidegger of the Greeks, on the basis of 'a fundamental experience of Being in poetry and thought' (pp. 15–16). Perhaps most importantly from our perspective, the Greek theory of *phusis* is fundamental enough to encompass not only the concept of nature (*natura*) but also everything to which that concept is traditionally opposed – position (*thesis*), soul (*psyche*), law (*nomos*) and even technics (*technē*) (p. 17). In the Greek world, *phusis* ultimately precedes the distinction between nature and technics: what now appear to us in the form of an opposition actually share a common origin.

It is clear, then, that *phusis* names the self-emergence of Being as such – an original and fundamental donation that includes both what subsequently came to be known as nature and technics – but before going any further we also need to clarify the precise *mode* of that emergence: *poiēsis* (*Hervorbringen*). As we have just seen, *phusis* names the coming-into-being of what is not yet a being and in 'The Question Concerning Technology' Heidegger cites Plato in the *Symposium* to argue that the Ancient Greek name for this process of presencing is *poiēsis*: 'Every occasion for whatever passes over and goes forward into presencing from that which is not presencing is *poiēsis*' (p. 10, Heidegger's translation). For Heidegger, that is to say, the Greek concept of *poiēsis* (a word which has subsequently become synonymous with 'manufacture' or 'production') articulates a more general process of disclosing or presencing that actually achieves its highest form of expression in *phusis*:

> It is of utmost importance that we think bringing-forth [*Hervorbringen*] in its full scope and at the same time in the sense in which the Greeks thought it. Not only handicraft manufacture, not only artistic and

> poetical bringing into appearance and concrete imagery, is a bringing forth, a *poiēsis*. *Phusis*, also, the arising of something from out of itself, is a bringing-forth, *poiēsis*. *Phusis* is indeed *poiēsis* in the highest sense. ('The Question Concerning Technology', p. 10)

If *phusis* is the 'highest' form of *poiēsis*, it is not because it is brought into appearance from outside by an efficient cause but because (as we already know from the famous discussion in Aristotle's *Physics*, 192b–193b), it contains within *itself* the source of its own bringing-forth: a flower – unlike a coat or a bedstead – blossoms without any outside agency or making. In the strict etymological sense of the word – which is obviously utterly alien to the modern biological concept popularised by Maturana and Valera – *phusis* is 'autopoietic': 'it' is the origin of its own presencing into beings.

For Heidegger, it is only from the vantage point of *phusis*' own poietic self-disclosure that we can begin to understand the Ancient Greek theory of *technē*. It is here that we begin to grasp the true audacity of Heideggerian philosophy of technology: *technē* is itself to be understood not as a mode of production or manufacture that brings something into existence out of inert matter (as the conventional reading of the history of the philosophy of technology would have it) but a mode of *poiēsis* that operates in accordance with *phusis*' own immanent process of poietic self-disclosure. As Rojcewicz argues, the theory of *technē* as poietic effectively seeks to level the opposition between the natural and the man-made, between the blossoming of a flower and the manufacture of a bedstead: '[a]ll making, all production, is natural, is self-emergence' (*The Gods and Technology*, p. 65). To start with, Heidegger's theory clearly requires a *destruktion* of the conventional reading of Aristotelian philosophy of technology as an efficient cause (*causa efficiens*) imposing form (*morphē*) upon matter (*hylē*) to his own end (*telos*): '*Aristotle's* Physics *is*', the philosopher famously argues, '*the hidden and therefore never adequately studied, foundational book of western philosophy*'.[7] Firstly, Heidegger rightly observes that what the Ancient Greeks understood by 'causality' (common to all of Aristotle's Four Causes) is not the comparatively modern idea of the efficient cause or maker with which it has become synonymous – such a concept actually appears nowhere in Aristotle's philosophy ('Question Concerning Technology', p. 8) – but a process of *poiēsis* that rhymes with, and assists, the *poiēsis* that is *already* underway in *phusis* itself:

> *Technē* is a mode of proceeding against *phusis*, though not yet so as to overpower it or exploit it, and above all not to turn use and

> calculation into principles, but, on the contrary, to retain the holding sway of *phusis* in unconcealment. (*Basic Questions of Philosophy*, pp. 179–80)

If *phusis* is already autopoietic, in other words, the role of the artisan or maker seems to be not to impose his own preconceived form (*morphē*) upon it for a particular end but to preserve, and be responsible for, this embryonic process of coming-into-appearance. Just as the artisan does not simply impress a pre-existing form upon inert matter, so what is called 'matter' (*hylē*) does not passively await the arrival of form: *hylē* (the original Greek word means 'wood' or 'forest' that is appropriate for woodwork rather than the abstract 'raw material' with which it has become synonymous) *already contains the potential* to disclose itself as a particular form. Far from the shape of a table being imposed upon a random piece of wood by a carpenter, as the conventional reception of Aristotle would have it, Heidegger argues that the 'very appropriateness of what is appropriated emerges more fully into view and reaches its fulfilment in the appearance of a table' ('On the Essence and Concept of *φύσις*', pp. 217–18). Perhaps the paradigmatic example of *technē qua poiēsis* in accordance with *phusis* is, of course, the work of art:

> The artist is a *technites*... because both the setting forth of works and the setting forth of equipment occur in a bringing forth and presenting that causes beings in the first place to come forward and be present in assuming an appearance [and] all this happens in the midst of the being that grows out of its own accord, *phusis*.[8]

In the famous re-narration of the Aristotelian theory of *technē* in 'The Question Concerning Technology', what would conventionally be depicted as the artisan's making or production of an artefact out of pure matter – a chalice fashioned out of silver – is transformed into the *co-creation* of all four (material, formal, final and the so-called efficient) of Aristotle's causes which, together and as an indivisible unity, are responsible for its poietic disclosure: 'The four ways of being responsible bring something into appearance. They let it come forth into presencing [*An-Wesen*]' ('Question Concerning Technology', p. 9).

In re-reading the Greek theory of *technē* so radically, Heidegger almost inevitably raises the question of *why* (if his reading holds force) this theory has been so totally forgotten. It is here, of course, that we come to his famous or notorious understanding of the history of western metaphysics as the reduction of the idea of Being as unconcealment or emergence to the modern scientific concept of a fixed, constant and present being. As we have already hinted at the beginning of this chapter, Heidegger does not see this as an essentially or exclusively *human* history: *Dasein* only exists (ek-sists, from *ek-stare*, stands out) in a relation to the truth of Being and thus Being is forgetting or reducing *itself* when it shows itself exclusively to *Dasein* in the form of present beings. What are the key episodes in this history? Firstly, Plato's theory of Forms – or rather the revelation of Being to which that theory conformed (p. 18) – evacuated Being's essence into a new supersensory *eidos* – a kind of eternal prototype or blueprint (*paradeigma*) – that pre-exists and determines the form all matter takes.[9] Such a theory represents a decisive break with the Pre-Socratic concept of Being in Heidegger's account: what was once the essence of Being itself – its self-emergence or unconcealment into beings – is now relegated to the degraded or inferior physical likeness of that (now metaphysicalised) essence. Now, if Platonism institutes a divide between what really is (*eidos*) and what is merely unformed matter awaiting its imprint (*hylē*), then Christianity – or what Nietzsche famously called Platonism for the people (*Introduction to Metaphysics*, p. 111) – reproduces that distinction as the opposition between uncreated being (*ens increatum*) and created beings (*ens creatum*). For Heidegger, Christian creationism intensifies the reduction of *phusis* to a hylomorphic act of efficient causality or making that is already at work within Greek metaphysics. Just as the Platonic architect locates the infinite *eidos* beneath the flux of matter and shapes it into form, so the Christian god pre-exists the world of his creation and brings it into existence according to his own predetermined design: it is almost as if the Greek theory of Being 'were cut to fit' the Christian worldview and interpretation of that which is as *ens creatum*.[10] Perhaps most radically, Christianity's concept of creation *ex nihilo* deprives being of even the meagre status of unformed *hylē* assigned to it by Platonism and leaves it in a total void or absence until the originary creative act takes place. Finally, Heidegger argues that the Christian concept of God as *ratio dei* – as a rational, thinking creator – in turn paves the way for the epoch of Deism and the post-Newtonian theory of the mechanised cosmos. If the universe is the product of an inherently rational design, Heidegger argues, then Being itself can be seen to become thinkable, calculable and manipulable in its essence and totality (*Introduction to Metaphysics*, p. 207). In this sense, Christianity unwittingly invents the modern biological concept of 'nature' as a physico-chemical object governed by invariant rules that are capable of being broken down, calculated and reassembled differently: Heidegger memorably argues in the

Contributions to Philosophy that 'machination (*Machenschraft*) contains the *Christian-biblical* interpretation of beings as *ens creatum* – regardless of whether this is taken in a religious or a secular way'.[11]

What, to pause momentarily, can we say of Heidegger's *destruktion* of the Ancient concept of *technē* before going any further? To begin with, we can easily see why he is such a canonical figure within twentieth-century philosophy of technology: Heidegger transforms the theory of technology from a mere regional or ontic science into nothing less than the theory (*thēoria*) of Being itself. It is perhaps surprising, then, that Heidegger's position as a thinker of what Derrida calls originary technicity remains in some dispute. As we have already suggested, Heidegger's ontological inflation of *technē* is routinely (if not always legitimately) charged with going hand in hand with a certain impoverishment of ontic or empirical technology which returns the latter to its traditional auxiliary or marginal position within ontology. For Derrida himself, Heidegger's claim that '*phusis*' – of all possible signifiers – is the privileged name for the originary disclosure of Being as such is arguably symptomatic of this implicit repression of *technē*:

> Heidegger, a thinker who is very attentive to the great question of *technē*, to the question of the relation between technics and philosophy, technics and metaphysics, technics and the West, perhaps remains, at a certain moment, tempted by a certain relegation of the technical to a secondary position in relation to a pretechnical originariness or a *phusis*.[12]

By identifying *phusis* – the word that will, of course, later come to be translated as 'nature' – with 'Being itself or the totality of being' (*Echographies of Television*, p. 133), Heidegger is deemed by Derrida and other critics to merely reproduce the metaphysical hierarchy of nature over technology at a new ontological level: 'a presence, a present, a presentifiable essence, a being as presence of *phusis*, would perhaps reconstitute itself, not simply before any technics in the modern sense, but before any *technē*' (p. 133). If this critique of a certain 'unexamined naturality'[13] at work in Heidegger's ontology is arguably somewhat crude – Derrida is, of course, well aware that *phusis* is not to be confused with any metaphysical determination of nature (p. 133) – the fact remains that *technē is* clearly placed in an ontologically secondary position by Heidegger: why else would the philosopher define it as 'a mode of proceeding *against phusis*' (emphasis mine)? Just as we may detect a residual ontological priority of *phusis* over *technē*, however, so there is also an implicit hierarchy at work within the specific mode of coming-into-appearance that they both allegedly share: *poiēsis*. On the one hand, *phusis* is defined as *poiēsis* in its 'highest' form: it is the originary paradigm of autopoietic bringing-forth. On the other hand, *technē* is necessarily relegated to a secondary or derivative mode of *poiēsis*: it merely follows and supports the primordial bringing-forth that is already underway in *phusis*. Finally, we might even go so far as to object to the poietic account of *technē* in its entirety, on the more empirical grounds that such a quasi-natural bringing-forth scarcely does justice to the diverse modes of technological production across the last 2500 years. Whilst *poiēsis* 'could coherently be applied to the forms of production known to the Greeks', Mark B. N. Hansen objects, 'it simply cannot be extended to cover the category of modern production': modern and postmodern technologies manifestly do not act in accordance with *phusis* (*Embodying* Technesis, p. 115). Perhaps such a claim over-extends itself – Heidegger usually (but not, as we will see, exclusively) associates the category of *poiēsis* with ancient technology alone ('The Question Concerning Technology', p. 14) – but what *all* modes of technology, ancient and modern, do share is that they are nothing more than the a-lethic disclosures of a primordial, non-technological *phusis*: ontic technologies are, in this sense, merely the passive carriers of an ontological donation. In our attempt to uncover the full extent of his commitment to an ontico-ontological difference that relegates ontic technologies to a revelation (*Offenbarung*) of a prior non-technological condition, we might come to wonder whether Heidegger is actually a thinker of originary technicity at all: what his philosophy presides over is less the technologisation of *phusis* than the ontological 'naturalisation' of technology.[14]

2

In order to better evaluate the legitimacy of the claim that Heidegger reduces ontic technology to the passive medium by which an ontological donation reveals itself, we need to turn from Being itself to the privileged recipient of that revelation: *Dasein*. It is in *Being and Time* that Heidegger articulates his existential analytic of *Dasein* and, revealingly, what Section 15 of that book calls *das Zeug* – stuff, tools or equipment – is the means by which *Dasein*'s fundamentally ontological mode of existence is disclosed to it.[15] As is very well known, Heidegger's discussion of equipmentality furnishes him with two key concepts to describe our interaction with our environment: *Vorhandenheit* (the present-to-hand) and *Zuhandenheit* (the ready-to-hand or, more simply, the 'available').

On the one hand, *Zuhandenheit* describes the pre-theoretical mode of existence in which we first encounter the world, where objects are experienced, not as things in themselves, but as part of a world of everyday practical concerns and involvements. On the other hand, *Vorhandenheit* names the theoretical or cognitive mode of knowledge which subsequently comes to see the world as consisting of individual things in themselves, standing on their own, entirely separately from *Dasein*. For Heidegger, of course, it is the hammer that serves as the heuristic device to explain the phenomenological priority of *Zuhandenheit* over *Vorhandenheit:* what starts out as something ready-to-hand – a pre-theoretical practical tool – only ends up becoming something present-to-hand – an isolated, theoretical object – after it breaks down (*Being and Time,* §15, p. 105). If *Dasein*'s hammer does not break down, however, then it never becomes an object in its own right but remains an invisible and indivisible part of the open-ended world of everyday involvements in which *Dasein* is unreflectively absorbed: I use my hammer in order to make a table and chair, for example, and I make a table and chair in order to work, or to eat dinner, and so on (p. 105). Perhaps understandably, Heidegger's account of the pre-theoretical relation between human and tool has attracted considerable interest from contemporary philosophy of technics. To what extent does the existential analytic acknowledge or repress the constitutive role of ontic technologies in the formation of *Dasein*?

To be sure, Heidegger's phenomenology of Being-in-the-World – which clearly refuses any residual Cartesian dualism between subject and object, consciousness and world, organism and environment – certainly *appears* to open up the possibility of the originary technicity of the human. It is not simply that *Dasein* divests human being of any zoological, biological or anthropological positivity but that this being is then wholly immersed in its factical environment: 'the destiny of the *who* (*Dasein*) is "tied" to intraworldly being, that is, to the *what*', argues Bernard Stiegler, 'it is what is "included" by its facticity' (*The Fault of Epimetheus*, p. 243). At the same time, however, a number of critics (particularly Derrida, Stiegler, Hansen and more recently Graham Harman) have once again detected a residual ambivalence towards technics within the existential analytic even at the very moment when it appears to be unequivocally embraced. Firstly, Derrida famously observes that (for all its critique of metaphysical humanism and anthropologism) Heidegger's analytic of *Dasein* remains complicit with a certain anthropocentric theory of *Zeug*: *Zuhandenheit* necessarily privileges *the hand tool* (handicraft; the hammer; the pen) over autonomised or industrialised forms of production as the authentic mode by which *Dasein* exists into relation with its world.[16] Now, Bernard Stiegler takes this critique a step further in order to insist that even those forms of technology that Heidegger deems authentic are stripped of all 'dynamic specificity' (*The Fault of Epimetheus*, p. 244) by their dependence on the hand. For Stiegler, Heidegger's account of *Zeug* reduces technics to little more than a supplemental role in the worlding of *Dasein*'s world: 'He always thinks tools as (merely) useful and instruments (merely) *as* tools' (p. 245). On the one hand, *Zeug* exists purely and simply *for Dasein*: the hammer is not an object in its own right, but is bound up, together with signs, in the referential totality of our involvements. On the other hand, *Dasein* alone, and not *Zeug*, is what discloses the world: the hammer's own unconcealing properties are immediately arrogated to *Dasein* itself. Perhaps most crucially, as Richard Beardsworth insists, this gesture unties the originary relation between the 'who' and the 'what' that characterises *Dasein* and assigns them to separate and antagonistic spheres: *Dasein* is re-anthropologised as an autonomous subject in search of its authentic mode of being and technics is expelled to the inauthentic realm of *das Man* (the They) (*Derrida and the Political*, pp. 151–2). In Stiegler's more detailed account, 'the *what* has no other dynamic than that of an inversion of the "authentic" dynamic of the *who*': *Zeug* 'will never have had the least *properly* unconcealing quality' (*The Fault of Epimetheus*, p. 244). Such a view of *Dasein* follows on logically from Heidegger's concept of Being as *phusis*: both Being's own self-revelation and the recipient of that revelation have at best only a contingent relation to technics.

For the embodiment theorist Mark B. N. Hansen, Heidegger's alleged blindspot regarding the ontic specificity of technics is the logical corollary of a larger idealist agenda within Heideggerian ontology and indeed modern philosophy of technology as such: what seems at first glance to be a critique of idealism in the name of technological embodiment, situatedness and wordliness is deemed by Hansen to preside over the ever more violent idealist reduction of technology's material exteriority that he names '*technesis*' (*Embodying* Technesis, pp. 1–24 and *passim*). It is this idealist reduction of technology, more than anything else, which explains why the German philosopher is unable to conceive of an 'in-itself' of the tool: the hammer exists *for Dasein* and has no meaning outside of that relation. As such, it is no coincidence for Hansen that the only way in which technological objectivity or materiality can be encountered at all is via the essentially theoretical mode of the *Vorhandenheit* – the present-at-hand hammer finally appears as an object in its own right rather than a mere extension of the hand only when it has broken down and interrupted *Dasein*'s immersion in its everyday activity. Yet, this acknowledgement of the autonomy of technology is at the cost of its necessary *idealisation*: we can only encounter our dependence on ontic technology theoretically in the form of an object that is a mere correlate to a thinking subject. To put it bluntly, then, Heidegger's theory of technology effectively ends up installing a phenomenological dualism that represses the autonomy of technology either way. Either technology is *Zuhandenheit* or *Vorhandenheit;* either it is instrument or idea, either it is assimilated wholly into *Dasein*'s own quest for authenticity or it appears as a theoretical object that stands over and against that quest for authenticity. If we are to break out of the dualism that afflicts Heidegger's account of *Zeug*, Hansen argues, then we need to explain what it excludes and Hansen's (surprisingly ontic) answer to that question is the phenomenological challenge posed by new media technology. Just as modern technology is not thinkable via the category of *poiēsis* (something that Heidegger himself, of course, would not contest), so it is not explicable according to the existential analytic. In Hansen's account, new media cut the Gordian knot between *Zuhandenheit* and *Vorhandenheit* because they represent a deep ontic condition for *Dasein* which is neither a ready-to-hand tool nor a present-to-hand object: new media technologies 'are implicated in our very Being-in-the-World as a central aspect of what enables experience as such' and so 'cannot be thematised in the same way as earlier technologies, that is, by being made explicit via the theoretical mode of knowing Heidegger calls *Vorhandenheit*' (*Embodying* Technesis, p. 105).

In the light of such undoubtedly powerful readings, Heidegger's theory of *Zeug* does appear to be symptomatic of a wider reduction of technological autonomy within his ontology but, as always with Heidegger, the story is never so simple. It is revealing that – for all their claims of a residual anti-technical idealism at the heart of the existential analytic that reduces technology to a mere supplement – neither Derrida nor Stiegler offer an ontically thicker account of technology as a constitutive force: Derrida's originary technicity (as we will see in Chapter 5) is not much less abstract and monolithic than *Zeug* whereas Stiegler's only example of what he calls 'dynamic specificity' is a form of technics that would scarcely be unfamiliar to Heidegger himself, namely, the work of art (*The Fault of Epimetheus*, p. 245). As we have just seen, Hansen is the only recent critic (albeit building on the work of Alfred Borgman and Avital Ronnell) to at least put a name to exactly *what* Heidegger is supposedly failing to think in his account of *Zeug*:

postmodern technologies like new media.[17] Nevertheless, in doing so, Hansen's critique of Heidegger's repression of ontic technology arguably risks re-imposing a Cartesian framework upon the existential analytic – subject and object, being and world, organism and environment – that Heidegger's account of Being-in-the-World is expressly designed to refuse. For Hansen, just as for Stiegler and Beardsworth, Heidegger's focus on the world-forming (*weltbildend*) power of *Dasein* in its ontological quest for authenticity ends up depriving *Zeug* of any autonomy: 'They [tools] owe their existence to the world-disclosing force of *Dasein*' (*Embodying* Technesis, p. 104). Now, I would not want to deny that there is a strong element within *Being and Time* that seeks to oppose the world-forming capacity of *Dasein* to the unyielding, technocratic world of *das Man*, but this cannot be at the expense of untying the 'already there' of which *Da-sein* itself is composed. By insisting that Heidegger arrogates all revelatory capacity to *Dasein* alone, Hansen risks reifying the *mutual* disclosure of human and world that takes place in the use of *Zeug* to the point where it becomes indistinguishable from the very hylomporphism Heidegger is attacking: 'the less we just stare at the hammer-Thing, and the more we seize hold of it and use it, the more primordial does our relationship to it become, and the more unveiledly is it encountered as that which it is – as equipment' (*Being and Time*, p. 98). If Heidegger does indeed deny the autonomy of the tool, it cannot be because it is merely the instrument of an independent, world-disclosing *Dasein* for, of course, *Dasein* can exist *nowhere* but in its world: it would be very unHeideggerian indeed to presume the existence of an anthropological '*Dasein*-in-itself' – which only voluntaristically enters into a relation to *Zeug* – as if Being-in-the-World were some kind of choice or accident. Just as Hansen risks performing the entification of *Dasein*, so he also risks presiding over an equally unHeideggerian positivisation of technology: there can be no technologies '*in themselves*' (*Embodying* Technesis, p. 104) any more than there can be a '*Dasein*-in-itself'. Perhaps in this respect Heidegger's non-objective, world-disclosive ready-to-hand hammer is a more radical example of originary technicity than we might first suspect: what Hansen sees as a pure instrument that plays no role in revealing *Dasein* to itself actually carries out the same pre-cognitive, non-thematisable conditioning function that he attributes to only the most advanced new media. Such a deep ontico-ontological reading of all technology might even tempt us to conclude that it is Hansen, rather than Heidegger, who could be accused of repressing the embodied technological real: what Hansen sees as the exclusive province of a very limited sphere of new media – namely, the ability to interface with the body at a pre-cognitive stage – is for Heidegger the condition of technology per se.

Why is it so difficult – despite so many attempts from Adorno to Hansen – to pin the charge of a simple anti-technological idealism, naturalism or romanticism on Heidegger's ontology? To start with, Heidegger himself is very clear that there is no question of his philosophy being simply 'for' or 'against' technics: 'I am not *against* technology. I have never spoken *against* technology, nor against the so-called demon of technology.'[18] It also goes without saying, though, that such claims have not stopped numerous theorists of technology from convicting fundamental ontology of a – variously psychoanalytic, philosophical or political blindspot – regarding ontic technics: Avital Ronnell goes so far as to speak of Heidegger's 'fear' of the technological and a distinctly psychoanalytic vocabulary of disavowal, repression and return dominates Beardsworth, Stiegler and Hansen's readings as well.[19] Despite the German philosopher's own careful attempts to demarcate *phusis* from nature, even Derrida claims to be able to detect 'a Heideggerian "pathos" ' at work in the thinking of *phusis* that 'remains antitechnological, originaristic, even ecologistic' (*Echographies of Television*, p. 133). Such alleged acts of repression or dogmatism are – by a predictable and not always entirely plausible magic – revealed to be the unconscious cause of the philosopher's disastrous ontic decision to ally himself with National Socialism in the Rectorate Address (see, for example, *The Fault of Epimetheus*, p. 207). Yet, as we have just discovered, there is an obvious danger in such attempts to critique Heidegger from a series of positions (whether psychoanalytic, historical materialist or even techno-positivist) which the philosopher himself is at pains to criticise of complicity with a metaphysics of subjectivity: post-Heideggerian philosophy of technology often appears to hypostatise ontological conditions into simple ontic objects or subjects in a way that is, if anything, conceptually pre-Heideggerian.[20] For me, it still remains possible to propose a critique of Heidegger's philosophy of technology as residually anti-technological, not on the reductive grounds that it expresses some kind of ontic 'preference' for nature over technology, so much as in the priority it grants a pre-technological ontology as the space in which such ontic technologies are primordially to be disclosed. By asserting the primacy of Being-as-*phusis*' poietic disclosure over being, for instance, we have seen that Heidegger effectively positions empirical technology as the ontically revealed rather than the ontological revealer: empirical technology is deprived of any role in its own hermeneutic disclosure. Perhaps the contemporary critique of Heidegger's philosophy of technics is at its most powerful when it contests the legitimacy of the ontico-ontological difference itself and the privilege it necessarily accords to the ontological gesture: what Heidegger calls ontology is – whether because of an originary state of empirico-transcendental contamination (Derrida); the mnemo-technical constitution of *Dasein*'s ontological temporalisation (Stiegler) or an embodied technological real (Hansen) – always already *given by* the ontic. If Heidegger insists that every ontic revelation (*Offenbarung*) can only appear against the backdrop of a pre-given ontological structure of revealability (*Offenbarkeit*), in other words, it becomes necessary to ask whether the ontic revelation itself might not in fact reveal (or at least participate in revealing) what is *then retroactively posited to be* the condition of its own revealability: the revelation reveals the revealer.[21] In one sense, we can already glimpse such an ontico-ontological aporia in the constitutive (rather than merely incidental) role Heidegger accords *das Zeug* in the existential analytic: it is the being *plus* the hammer that discloses the world in which both together reside.

3

In a larger sense, of course, it is around the later Heidegger – and particularly classic essays such as 'The Question Concerning Technology' (*Die Frage nach der Technik*) – that the contemporary debate about the ontologisation of technology is truly joined. To many readers, the philosopher's later account of technics – with its vocabulary of the 'essence' of technology, *das Gestell*, the danger, the saving power and the piety of thinking – will be so familiar that it hardly requires further rehearsal. Firstly, of course, we find Heiddegger's most succinct (if still arguably cryptic and misunderstood) formulation of the ontological reading of technology: 'Technology (*der Technik*) is not equivalent to the essence (*Wesen*) of technology' ('The Question Concerning Technology', p. 4). It is once again a question of understanding *technē* – not ontically – but as a distinctive mode by which Being shows itself as itself. However, it is now clearer than ever that there is a fundamental qualitative difference between the ways in which Being discloses itself in ancient and modern technologies. On the one hand, of course, ancient technology remains a mode of *poiēsis* (of bringing-forth) that operates in accordance with the self-emergence of *phusis* itself. On the other hand, though, modern technology is a mode of *Herausfordern* (challenging-forth) that 'puts to nature the unreasonable demand that it supply energy that can be extracted and stored as such' (p. 14). If the ancient silversmith understands his labour as carrying on a process of unconcealment that is already underway in *phusis*, the modern scientist sees himself as efficiently imposing his own rational *morphe* upon a *phusis* reduced to mere matter. For Heidegger, *Herausfordern* – far from operating in harmony with *phusis* like its ancient predecessor – performs a violence upon a *phusis* which is now disclosed to be not the 'emerging-abiding sway' of Being itself but nothing more than a standing reserve (*Bestand*) or resource of potential energy to be exploited, accumulated and distributed:

> The revealing that rules throughout modern technology has the character of a setting-upon [imposition or *Bestellen*], in the sense of a challenging-forth. That challenging happens in that the energy concealed in nature is unlocked, what is unlocked is transformed, what is transformed is stored up, what is stored up is, in turn, distributed, and what is distributed is switched about ever anew. Unlocking, transforming, storing, distributing and switching about are ways of revealing. (p. 16)

It will be no surprise by now to learn that Heidegger considers modern technology to be no more a human accomplishment than its ancient equivalent: both are the product of a prior ontological revelation (p. 18). As the philosopher goes on to make very clear, what appears to install the human in a position of absolute mastery over the natural world is once again something that has been donated *to* humanity: we are only able to see Being as mere natural raw material ready for exploitation by modern technology because Being has already shown itself to us in this form. Even the apparently rigorous empiricism of the modern physical sciences – which is commonly thought to pave the way for modern technology – itself presupposes the revelation of Being that is *das Gestell*.[22] Not only do human beings lack control of the revelation of Being at work within modern technology, but we are ourselves profoundly subject to that revelation: humanity itself is challenged – revealed as a particular kind of standing reserve of energy to be exploited – to, in turn, challenge nature. Such a revelation of Being that challenges man to challenge nature is what Heidegger calls *das Gestell*:

> Enframing (*das Gestell*) means the gathering together of that setting-upon which sets upon man, i.e., challenges him forth, to reveal the real, in the mode of ordering, as standing-reserve. Enframing means that way of revealing which holds sway in the essence of modern technology and which is itself nothing technological. (p. 20)

If *das Gestell* is the non-technological mode of revealing that makes modern technology possible, though, the philosopher dramatically declares that its sway also represents a state of danger, even of the supreme danger (p. 26). For Heidegger, characteristically, such a danger does not primarily consist in the threat to the human species or the planet posed by ontic technologies such as nuclear weaponry – modern technology is not dangerous in itself (p. 28) – so much as in the threat posed by *das Gestell* to the *ontological* relation between humanity and Being. On the one hand, *Dasein* is so thoroughly in the grip of the illusion of absolute sovereignty over Being conjured up by *das Gestell* that he is no longer able to recognise that it is still a mode of revealing of which he is merely the recipient: 'he does not apprehend Enframing as a claim... he fails to see himself as the one spoken to, and hence also fails in every way to hear in what respect he ek-sists, from out of his essence, in the realm of an exhortation or address' (p. 27). On the other hand, *das Gestell*'s total monopoly over the process of unconcealment blocks off all other possible revelations of Being including, of course, *poiēsis*: 'Where this ordering holds sway, it drives out every other possibility of revealing. Above all, Enframing conceals that revealing which, in the sense of *poiēsis*, lets what presences come forth into appearance' (p. 27). In the end, *das Gestell* not only excludes all other possible revelations but it even occults its own status *as* a revelation and the ontological structure of Being as a process *of* revealing: enframing simply becomes synonymous with 'what is' (p. 27).

To many eyes, the later Heidegger's *destruktion* of modern technology represents the logical conclusion of the ontologisation of technology we have witnessed throughout his work from *Being and Time* onwards. Only by thinking the ontological essence of technology – the fact that it remains, for all its occultation, a sent disclosure of Being – can we begin to understand empirical technology itself. It is precisely this attempt to oppose the essence of technics to technics itself, however, that has led to the charge that Heidegger's ontology presides over a reduction or exclusion of technics from the proper domain of thought. As is now well documented, Derrida (and later Stiegler) argue that this opposition protects metaphysically 'the possibility of a thought that questions... from any original and essential contamination by technology'. Both would insist that the allegedly pre-technological thinkable 'essence' of technology must in some sense already be given by, with and through, technics (*Of Spirit*, p. 10; *The Fault of Epimetheus*, pp. 14, 18). Yet, once again, we must be very careful not to jump to what would still be distinctly pre-Heideggerian conclusions here: Heidegger's speaking of a pre-technological 'essence' (*Wesen*) does not automatically commit him to idealism any more than speaking of *phusis* commits him to naturalism. For Heidegger, *das Gestell* is precisely *not* a universal genus (*quidditas*) or abstract metaphysical concept (*eidos, idea*) under which all individual technologies are to be subsumed[23] but rather the self-granting or bestowal (*Gewähren*), of Being itself to *Dasein*: 'Every destining of revealing comes to pass from out of a granting and as such a granting' ('The Question Concerning Technology', p. 32). If the essence of technology is clearly not a Platonic *eidos* that pre-exists all empirical technology (p. 31), though, Derrida is right to say that it does still *precede* ontic technology in the way that an ontological process of revealing absolutely precedes the particular revelation: what is prohibited by the ontological gesture, as we have already observed, is the possibility that ontic technology itself may participate in the questioning of its own essence. Perhaps modern technology (for all its alleged autonomy) becomes even *less* revelatory than its ancient equivalent when viewed from this ontological vantage point: whereas the hand tool plays a constitutive role in *Dasein*'s world-forming – recall the hammer in *Being and Time* – the airplane is entirely dependent upon the prior 'ordering of the orderable' (p. 17) that is *das Gestell*. Such a gesture can only risk reproducing the metaphysical version of Aristotelian philosophy of technology at the level of the ontico-ontological difference. Just as Aristotle's *Physics* (at least according to the conventional reading) consigns the tool to a state of total inertia until the arrival of the efficient cause, so Heidegger insists that the coming-into-appearance of the machine remains wholly contingent upon the disclosive power of a pre-technological agency: 'technological activity... always merely responds to the challenge of Enframing, but it never comprises Enframing itself or brings it about' (pp. 20–1). By reducing ontic technology to the effect of a prior ontological revelation, Heidegger risks elevating Enframing to the ontological equivalent of Aristotle's mythical efficient cause (see also *Embodying* Technesis, p. 113).

For Heidegger, of course, it is precisely the fact that *das Gestell* is a revelation bestowed *upon* man, rather than one created or imposed *by* him, that is the 'saving power' that lies at the heart of the supreme danger of modern technology. It is only by thinking the 'ambiguous essence' (p. 33) of technology as a disclosure granted to *Dasein* by Being, rather than a technological object to be disposed of according to the will of a subject, that *Dasein* is able to receive the possibility of a 'free relation' (p. 3) to modern technology. Yet, in the light of the preceding argument, we might begin to wonder whether Heidegger's position here in fact risks circularity: it is only because he has *already presupposed* that the essence of technology is nothing technological that such a free relation to technology becomes possible in the first place. To put it bluntly, Heidegger's belief in what he calls 'the piety of thought' (p. 35) as a means of decisively confronting *das Gestell* risks turning into something of a self-fulfilling prophecy. By taking the decision to oppose Being and technology at the outset via the mechanism of the ontico-ontological difference, the philosopher effectively creates the very pre-technological point of origin that *Dasein*'s pious act of *anamnesis* claims to recollect. Perhaps one privileged sign of Heidegger's residual anti-technological bias is that the old category of *poiēsis* or bringing-forth is re-invoked at precisely this moment in the essay to describe the 'origin' of *das Gestell* in a sent disclosure of Being:

> Revealing is that destining which, ever suddenly and inexplicably to all thinking, apportions itself into the revealing that brings forth and that also challenges, and which allots itself to man. The challenging-revealing has its origin as a destining in bringing-forth [*Hervorbringen*]. (pp. 29–30)

Why, after constantly arguing that *das Gestell* occludes *poiēsis*, does Heidegger now see it as *itself* poietic in origin? If *poiēsis* has been more or less exclusively associated with ancient technology up to this point, it now seems to have become the proper name not simply of one *particular* mode of revealing but of *the process of revealing as such*. Both ancient and modern technology – even thought the latter effectively represents the final occultation of the poietic mode – are outworkings of the self-emerging of *phusis* that is *poiēsis*. Such a remarkable extension of the concept of *poiēsis* to cover *das Gestell* as well means, of course, that *phusis* – which has already been awarded the distinction of being *poiēsis* in its highest form – is re-introduced as the measuring stick by which modern technology is to be judged and, all too inevitably, found wanting: 'Enframing, in a way characteristic of a destining, blocks *poiēsis*' (p. 30). In this context, contemporary technology is reduced to nothing more than a (modern) derivation of an (ancient) derivation of *phusis*' own autopoietic form of presencing: *das Gestell* is a form of technical *poiēsis* that does not merely 'proceed against *phusis*' (*Basic Questions of Philosophy*, pp. 179–80) in the manner of its ancient equivalent but against its *own* origin as poietic.

Who, finally, are the 'we' who are constantly invoked throughout Heidegger's essay? It is the answer to this question, more than anything else, which reveals what Agamben calls the anthropological machine within the later Heidegger's philosophy of technology.[24] As we have seen throughout this chapter, technics for Heidegger is never simply a human accomplishment, so much as a self-revealing (or self-withholding or occultation) of Being itself *to* humans. However, we have also seen that man is still accorded a privileged position over all other beings within *das Gestell* as the only genuine recipient of Being's revelation: 'Only to the extent that man for his part is already challenged to exploit the energies of nature can this ordering revealing happen' (p. 18). Not only does man challenge nature to reveal itself as energy, but he *himself* is challenged by Being to put his own energy into carrying out this action; phrases such as 'man power', 'labour supply' or 'human resources' eloquently testify to the fact that humans are just as much subject to *Herausfordern* as nature. To Heidegger, of course, *Dasein* is uniquely receptive to the self-disclosure of Being – after all Being needs something to disclose itself to – and this attunement not only makes humanity peculiarly vulnerable to the prevailing disclosure but, intriguingly, also seems to grant it at least the possibility of a certain freedom *from* that disclosure:

> Yet precisely because man is challenged more originally than are the energies of nature, i.e., into the process of ordering, he never is transformed into mere standing-reserve (*Bestand*). Since man drives technology forward, he takes part in ordering as a way of revealing. (p. 18)

By participating in the very process of ordering nature – as the very being who is challenged to accomplish it – man can never become simply a passive object of that ordering, but always remains a being who, as Rojcewicz observes, 'retain[s] an awareness (or at least the possibility of an awareness) of the demands that are being placed upon them' (*The Gods and Technology*, p. 96). Just as *Dasein* is the only being whose Being is an issue for it – which is why it can never be wholly lost in the inauthentic world of the They (*das Man*) – so humanity's essentially ontological mode of being means that it is alone amongst natural beings in remaining immune to *Herausfordern* despite (or even because of) the fact that it drives that process forward. Quite simply, we can never lose the potential to see that modern technology is a sent disclosure of a pre-technological Being – a 'way of revealing' – because *we are nothing other than* this potential. Such a remarkable claim is, in many ways, merely a logical corollary of Heidegger's ontological reading of technology – which presupposes the very possibility of just such a free relationship to technology because, as we have just seen, it only ever sees the latter as a revelation of a thinkable pre-technical essence – but, even so, it is not difficult to see why the identification of *Dasein* as the embodiment of that freedom has prompted allegations of anthropocentrism from post-Heideggerian philosophy of technology. If *Dasein*'s unique capacity for ontological questioning enables it to free itself from the grip of *das Gestell*, Hansen argues, then Heidegger is able to secure 'nothing less than the categorical triumph of the human over any merely external interference with the call of Being' (*Embodying* Technesis, p. 121). Perhaps this charge is once again a little exaggerated – we have already seen how the existential analytic expressly forbids any categorical distinction between subject and object so no such humanist 'triumph' over the external world of facticity is possible – but it is certainly tempting to suspect that *Dasein* here occupies an analogously privileged position to the *anthropos* of metaphysical anthropology. For Derrida, as his own career-long deconstruction of Heideggerian ontology makes clear, the thinking of Being remains – despite its allegedly pre-anthropological status – a 'thinking *of* man' in both senses of the genitive: what inspires the critique of anthropology in such texts as the 'Letter on Humanism', Derrida insists, is scarcely a blanket rejection of the name of 'man' so much as an attempt to restore his proper dignity or essence (*das Wesen der Menschen*) to himself.[25] In a gesture that we will see recurring in the work of Derrida and Stiegler in Chapters 5 and 6, Heidegger's ostensibly de-anthropologising definition of *Dasein* as ek-sistence – as the only being who properly stands in a questioning relation to Being – permanently risks re-anthropologising it *as precisely and uniquely this relation*: 'we' are the only being who asks the ontological question, the questioning being, the being *of* the question ('The Ends of Man', p. 135).

4

In drawing this chapter to a close, however, I would briefly like to turn to what – if my critique has any purchase – might lie on the other side of Heidegger's ontological reading of technology. What exactly is being excluded or repressed by the philosopher's reduction of technics to a destined disclosure of Being to *Dasein*? To be sure, continental philosophy of technology has not exactly been slow to respond to this question and we now know the thrust of its answer: Heidegger's ontology perpetuates an idealist anthropologisation of technics that reduces it back down to a question *for* a thinking *Dasein*. Such an anthropological machine has been observed at work in everything from the German philosopher's instrumentalisation of techno-science on the one extreme (Derrida, Stiegler, Badiou) to his entification of the world-impoverished animal on the other (Derrida, Agamben). Perhaps the most radical recent critique of Heidegger's idealism is that – for all its questioning of the illusory sovereignty of the human over the natural world – the later Heidegger's revelation of the mutual belonging together of Being and *Dasein* in *das Ereignis* ultimately forbids the existence of any world that is independent of the thinking subject: Quentin Meillassoux places Heidegger within the tradition of idealist 'correlationism' that begins with Kant and (allegedly at least) culminates in the quasi-theological phenomenology of a Levinas or Derrida.[26] For all the major differences between and within this array of thinkers, however, it is possible to argue that in one sense the ground of their critique remains largely the same: what Heidegger terms the proper ontological sphere of *Dasein*, alone, is actually carved out – by a literal and violent philosophical de-cision – from a pre-given complex and inhuman world.

It may be that we can find one final (if strangely neglected) symptom of Heidegger's marginalisation of technics in his philosophy of science. As is now well documented, 'The Question Concerning Technology' was not originally a free-standing piece of work but just one in a series of seven lectures by philosophers and scientists given to the Bavarian Academy of Fine Arts as part of a 1953 symposium on the theme of 'The Arts in the Technological Age'. During the symposium, Heidegger's own lecture was immediately preceded by a contribution from the renowned quantum physicist Werner Heisenberg entitled 'The Picture of Nature in Contemporary Physics'. For Heisenberg, whose essay was largely a summary of the revolutionary discoveries that had transformed physics over the previous 30 years, what the quantum revolution had taught us was that the 'picture of nature' (*Das Naturbild)* given by contemporary physics was just that – a picture. Natural science could no longer claim to represent with any certainty objects or laws in the natural world that exist independently of the act of observation. Such, at least, was the radical conclusion Heisenberg had arrived at some 30 years earlier when he formulated his famous Indeterminacy Principle: what is being recorded by scientific experiments on quantum phenomena is not a description of the natural world as it really exists but a result that is at least partly produced by the experiment itself.[27] If Newtonian physics were able to describe certain universal laws of causality that governed all phenomena – every moving object in time and space must possess both a position and a momentum, for example, regardless of whether these are being measured or not – the Heisenberg-Bohr Copenhagen school of quantum physics judged it a priori impossible to ascribe properties to quanta prior to the act of collapsing the wave function through measurement: it is simply meaningless, they argued, to speak of the position or momentum of a particle 'in itself' before the theoretical gamma ray microscope arrives on the scene. In 'The Picture of Nature in Contemporary Physics', Heisenberg describes how modern physicists have largely surrendered any claim to ontological realism as the necessary price for producing reliable, predictable and applicable data from their experiments:

> To this extent, we have renounced the type of description of nature that was customary for centuries and that has been valid as the self-evident goal of all exact natural science. Even provisionally, we cannot say more than that in the field of modern physics, we have resigned ourselves, and we have done so because our representations are dependable.[28]

For Heidegger – whose 'Question Concerning Technology' twice alludes to Heisenberg's own lecture – the quantum revolution is by no means something merely 'dictated by a commission of scientists', of course, but rather the logical and historical conclusion of the modern technological mode of revealing he calls *das Gestell*. 'It is challenged-forth by the rule of Enframing, which demands that nature be orderable as standing-reserve' ('The Question Concerning Technology', p. 23). As we have already seen, Heidegger argues that the epoch of modern science presupposes *das Gestell* rather than the other way around: it is only because Being *has already revealed itself* in the form of a set of calculable objects standing ready to be exploited that the scientific experiment is able to observe it as such. Not only is the essence of technology nothing technical but, as the philosopher argues in another of his quietly astonishing claims, 'Physics itself is not the possible object of a physical experiment' ('Science and Reflection', p. 176). To Heidegger's way of thinking, the quantum revolution described by Heisenberg raises this act of scientific *Herausfordern* to a radical new level: modern physics has imposed itself so completely upon *phusis* that the latter has entirely disappeared from view and all that remains visible to us is our own act of imposition. Whilst modern physics is content to renounce the existence of an observer-independent order reality, the philosopher insists that the one thing it can never renounce is the demand that 'nature reports itself in some way or other that is identifiable through calculation, and that it remains orderable as a system of information' ('The Question Concerning Technology', p. 23). If the Ancient Greeks saw causality as an act of *poiēsis*, and the modern scientist reduced it to an act of making of efficient causality, the quantum physicist shrinks causality still further into the successive recording of data with no sense of a real or necessary connection between such recordings: a light particle does not, strictly speaking, even exist between one act of measurement and the next. Perhaps the most famous example of this eviscerated theory of causality can once again be found in Heisenberg's Indeterminacy Principle. Whereas classical physics had shown that, if the precise position and velocity of a particle is known at any given moment, then we can determine its position and velocity in the future, quantum physics insists that, because we are unable to measure such data with any precision, we are left with an irreducible range of future possibilities – of superpositions – that can be, at best, probabilistically determined. In Heisenberg's dramatic words, 'quantum mechanics establishes the final failure of causality' (*'Über quantentheoretische Umdeutung kinematischer und mechanischer Beziehungen'*, p. 83). Such a total abandonment of *phusis* in the name of greater mathematical formality represents, for Heidegger, the apotheosis of the technological will to mastery over the natural world:

> In this way the illusion comes to prevail that everything man encounters exists only insofar as it is his construct. This illusion gives rise in turn to one final delusion: it seems as though man everywhere and always encounters only himself. Heisenberg has with complete correctness pointed out that the actual must present itself to contemporary man in this way. *In truth, however, precisely nowhere does man today any longer encounter himself, i.e., his essence.* ('The Question Concerning Technology', p. 27)

What, with the undoubted luxury of more than 50 years of hindsight, can we now make of Heidegger's critique of quantum physics? It is perhaps tempting to suspect that, if the Indeterminacy Principle did not already exist, it would have been necessary for Heidegger to invent it, for it appears so totally to vindicate his ontological reading of techno-science. After all, a world where the only objects the physicist can encounter are the products of his own acts of observation is indeed a world in which man 'everywhere and always encounters only himself'. Yet it is also possible to wonder whether Heisenberg's Demon also represents something of a Heideggerian *Deus ex machina* in this context, which is all too conveniently summoned to earth to pass divine judgement upon the vicissitudes of *das Gestell*. To begin with, Heidegger's claim that quantum physics, like all science, still demands that 'nature reports itself in some way or other that is identifiable through calculation' (p. 23) not only attributes to the quantum revolution an ontological pretension that Heisenberg specifically refuses – there is no longer any pre-existing 'nature' of which science can make demands – but also risks exaggerating its mathematical precision. Put simply, what is at stake in the Indeterminacy Principle is not the ever-greater reduction of the incalculable *to* the calculable, as Heidegger claims, but rather the discovery of an irreducible incalculability *by* calculation itself. Such is the infinite possibility of superposition inherent in the wave function that, whilst a position can indeed be mapped probabilistically to an extraordinarily reliable degree, it can never be predicted with 100 per cent accuracy. Secondly, we might argue that the subsequent evolution of quantum theory – and particularly the attempt to resolve some of the epistemological paradoxes thrown up by the pure formalism of the Copenhagen school[29] – also militates against Heidegger's attempt to recruit that theory to the ontological cause of *das Gestell*. For Hugh Everett III, renowned pioneer of the so-called 'Many Worlds' interpretation of quantum theory, the Indeterminacy Principle – far from making reality dependent upon the existence of an observer – actually describes an entirely objective and observer-free reality albeit one which consists of an infinite number of parallel worlds. Whilst Heisenberg argues that each and every quantum possibility only becomes actualised by the collapsing of the wave function into a single state – via the measuring of, for example, an electron's position by an observer – the Many Worlds school sees every possible quantum state as *actually co-existing* in different universes.[30] If the Copenhagen school leaves us with a set of probabilities that cannot be calculated definitely, the Many Worlds interpretation opens up a set of *actualities* whose coming-to-be clearly cannot be probabilistically predicted in any classical sense of the term: what classical physics would call 'probabilities' have always already been actualised somewhere in the multi-verse. Perhaps we might still be tempted to see the evolution of the Many Worlds interpretation as yet another vindication of Heidegger's critique of quantum theory – in the sense that it explodes the illusion that human beings construct their own *phusis*, rather than are granted a *phusis* that always exceeds their grasp – but it only takes a moment to realise that the opposite is closer to the truth: it is the ever-greater process of calculation and objectification made possible by contemporary physics – rather than some pre-technological donation of Being – which is generating a deeper and more fundamental incalculability, an objectlessness at the heart of every object. In other words, it is technics *itself* that performs the very critique of calculative or instrumentalist reason that Heidegger claims can only be accomplished by the non-technical essence of technology.[31]

5

What, to conclude, is at stake in Heidegger's ontological theory of technology? To start with, I would argue that Heidegger is undoubtedly the most radical philosopher of originary technicity we have encountered thus far in this book: what the philosopher calls the 'essence' of modern technology is neither a historical moment to be dialecticised (Marx); not a metaphor for mind (Freud) nor even an embodiment for the symbolic constitution of the subject (Lacan) but nothing less than an ontological condition. However, it is perversely this very *ontologisation* of technology – its elevation from a contingent factical phenomenon into something that is coterminous with the very fate of the west as such – that arguably betrays an anti-technological residue within Heideggerian ontology. For Heidegger, such an ontological inflation of technics is paid for by a corresponding reduction of ontic technologies to the second-order revelations of a pre-technological, poietic and *phusis*-dominated essence. If Marx, Freud and Lacan implicitly appeal to a position outside the immanent processes and organisations of technics itself – whether it be collective unalienated labour or a psyche capable of grasping the extent of its own machinic status – so in the same way Heidegger presupposes that (however all-pervading it may be) our technological condition must still remain thinkable from some implied pre-technological vantage point. Such is the basis of Jacques Derrida's critique of the philosopher's attempt to posit an ontological essence of technology: 'Isn't he tempted to subtract, in this way, the thinkable or thinking from the field of technics? Doesn't he suggest that there is a thinking pure of all technics?' (*Echographies of Television*, p. 133). Perhaps (although here we must leave behind any orthodox Heideggerianism) it might only be by affirming such an ontico-ontological *aporia* of technics – rather than opposing ontic technology to some prior ontological outside – that we can begin to question the all-encompassing instrumentalisation of *das Gestell*: what the quantum revolution shows is that technics itself (rather than some allegedly extra-technical poietic essence) can accomplish the de-essentialising, de-instrumentalising and de-totalising task Heidegger always accords to Being. In the next chapter, we will explore how Derrida's deconstruction seeks to pursue precisely this originary contamination of technics and the essence of technics in his own theory of originary technicity.

5
The Other

> There is no deconstruction which . . . does not begin . . . by calling into question the dissociation between thought and technology . . . however secret, subtle, sublime or denied it may be.[1]

In the thought of Jacques Derrida, the theory of originary technicity we have been tracking throughout this book obtains its most explicit formulation. It is Derrida – more than Marx, Freud, Lacan or even Heidegger – who, I will argue, enables us to glimpse the full extent of what the thinker himself revealingly calls *'our technological condition'*.[2] As we have already noted in Chapter 1, Derrida's interest in an 'originary' concept of technics is apparent from the very beginning of his work on *arche*-writing: the philosopher famously calls for a new exploration of the 'question of *technology*' which would exceed the classical opposition between 'life and death' as early as his 1966 lecture on psychoanalysis ('Freud and the Scene of Writing', p. 228, Derrida's emphasis). Such interest only increases as his philosophical itinerary broadens to encompass a whole new wave of writing machines: photography, television, the computer and so on.

For Derrida, as the philosopher himself puts it in the epigraph to this chapter, what is called 'deconstruction' *always* begins with the deconstruction of the opposition between thought and technology. Such a radical and generalised concept of 'technics' – as one, albeit privileged, name for the logic of spatio-temporal difference, repetition and radical alterity with which deconstruction is famously concerned – emerges in his thought as the basic and irreducible condition of nature, matter, ideality, thought, perception, mnemics, affect – in short the entire sphere of 'the living'. If deconstruction thus remains the most self-conscious philosophy of originary technicity we possess – inasmuch as it destroys any concept of a pure, natural or non-technical point of origin – Derrida's philosophy has nonetheless been subject to consistent charges that, like Heideggerian ontology before it, it, too, contains a residual idealism: Bernard Stiegler, Richard Beardsworth and, most recently, Mark B. N. Hansen have (albeit in different ways) argued that – even despite the philosopher's apparently exhaustive work on the body, the animal and the technical – the deconstruction of the opposition between thought and technology too often seems to be carried out by thought alone. Perhaps the most visible and controversial evidence of this (alleged) idealism at the heart of deconstruction concerns the later Derrida's so-called 'religious' turn where, it is argued, the philosopher moves away from the quasi-empirical vocabulary of his earlier work on writing towards a quasi-transcendental vocabulary drawn from the Judaeo-Christian tradition: the promise, the impossible, negative theology and the messianic.

This chapter explores Derrida's career-long deconstruction of the relation between thought and technics from his articulation of an originary writing that underlies the living presence of speech in seminal early texts like *Of Grammatology* (1967), through his account of the virtuality or mediation that inhabits the phenomenon of absolute spatio-temporal immediacy in works like *Echographies of Television* (1986) and *On Touching – Jean-Luc Nancy* (2002), all the way up to the alleged 'religious' turn within his later thought. To what extent does Derrida's theory of originary technicity give us a new means of access to the constitutive role played by technology *within* thought or does it perpetuate, even deepen, the exclusion of technology *from* thought to the point where it becomes a mere object for prior philosophical inquiry?

1

In his early triptych *Voice and Phenomena*, *Of Grammatology* and *Writing and Difference* (1967), Derrida begins to formulate his account of the originary technicity of life via the famous concept of *arche*-writing (*arche-écriture*). It is the later Husserl's theory of transcendental historicity that provides the early Derrida with his first point of access to the question of a generalised technicity that exceeds all metaphysical oppositions between life and death, the transcendental and the empirical, ideality and materiality and so on. As Husserl famously argues, transcendental phenomenology seeks to reduce or bracket off facticity – the world in all its changeability and contingency (*Verkörperung*) – in order to gain access to ideal objects. However, what makes the phenomenological reduction of the world possible in the first place – and thus access to ideality – is, as Derrida shows, nothing other than the factical act of *writing*.[3] To briefly summarise his reading, Derrida puts his finger on an significant moment in Husserl's essay 'The Origin of Geometry' where the latter acknowledges that it is only because Euclid and the other early geometers *wrote down* their principles – in repeatable graphic form – that geometry ceased to be just a subjective idea in their own consciousness and was transformed into an objective, universal science. Now, if this is the case, then geometrical objects occupy a curious middle position between the ideal and the material, the transcendental and the factical. On the one hand, geometrical objects are obviously ideal: they transcend time and space in order to exist in exactly the same way for everyone across history. On the other hand, though, geometrical objects are also material: they must have been inscribed in a particular time and space in order to transcend anything in the first place. For Derrida, this aporetic or undecidable relation between the empirical and the transcendental – where the ideal *requires* the material in order to become itself in the first place – arguably represents the beginning of his career-long attempt to articulate the immanent relation between technics and thought. If writing makes possible the phenomenological reduction, and the reduction, in turn, is what gives access to the ideal, then this means that technics is already inside the ideal as its condition of possibility: writing, to put it another way, is *irreducible*. Such are the 'origins' of Derrida's own version of originary technicity.[4]

It is, of course, possible to detect an equivalent re-thinking of the history of the empirico-transcendental opposition in terms of a quasi-transcendental logic of originary technicity throughout the early phase of Derrida's work. As Richard Beardsworth amongst others has argued, the repression of technics (usually, but not exclusively, thought under the sign of writing) from its constitutive role within the formation of ideality becomes an important symptom for any metaphysics of presence. To recall Chapter 1 of this book, Derrida begins by problematising Plato's foundational gesture of valorising a living, immortal *anamnēsis* over the artificial or technical memory that is *hypomnēsis*: 'Memory is . . . contaminated with its first substitute: *hypomnēsis*' ('Plato's Pharmacy', p. 111). By the same token, Jean-Jacques Rousseau and his modern anthropological disciple Claude Levi-Strauss posit a pre-social state of nature into which culture, language and, particularly, writing only *subsequently* arrive but, once again, Derrida is able to argue that the violence of the letter is always already at work (*Of Grammatology*,

p. 110). For Sigmund Freud, as we saw in Chapter 3, it becomes clear that the scriptural metaphor for the psychic apparatus conceals what Derrida reveals to be a more generalised homology between what are, in effect, merely two different species of writing machine. Perhaps most definingly, Ferdinand de Saussure's *Course in General Linguistics* (1917) – for all its scientific pretensions – seeks to re-establish an entirely Platonic hierarchy between the spoken and the written signifier, but the Swiss linguist's theory of language as a series of differences without positive terms compels him to conclude, in spite of himself, that *all* language belongs to a general category of 'writing' (*Of Grammatology*, pp. 27–64). If Derrida is able to demonstrate the existence of an originary writing, materiality or technicity within any philosophical system – which is to say the logical or chronological priority of something that, at the same time, remains essentially extrinsic, posterior or supplemental – then the philosopher is able to deconstruct the metaphysics of a pure, singular or original presence on which such systems found themselves: the origin of metaphysics is revealed to be not at one with itself. In the classic essay 'Différance', for instance, *technē* becomes one of the privileged signifiers of the non-identity of Being itself: *différance* is 'culture as nature different and deferred, differing-deferring: all the others of *phusis* – *technē, nomos, thesis*, society, freedom, history, mind, etc... *Physis* in *différance*'.[5]

For the early Derrida – writing, as he is, in the wake of both post-Husserlian and Heideggerian phenomenology and the structuralist revolution in the so-called human sciences (*les sciences humaines*) – it is arguably within the field of *anthropology* that the question of technics can best be witnessed. Originary technicity makes it impossible to define the human either transcendentally (as a soul, mind or *ego cogito*) or empirically (as a historical, material or biological entity) because whatever intrinsic nature we allegedly possess is constituted in its very interiority by what it positions 'outside' itself. According to the famous logic that is first developed in the context of his reading of Rousseau in the *Grammatology*, what apparently serves to *supplement* the state of nature – culture, language, writing and the entire field of technics – actually exposes an originary *lack* within what should be the integrity or plenitude of the human being itself which calls for supplementation. To put it in Derrida's own words, what we call the 'human' cannot be said to pre-exist its supplements – in the way that, say, an Aristotelian efficient cause precedes its instruments – but is rather *constructed retroactively* on the basis of this process of supplementation: 'Man allows himself to be announced to himself after the fact of supplementarity, which is thus not an attribute – essential or accidental – of man' (*Of Grammatology*, p. 244). If the identity of the human is always bound up with its non-human supplements, then the early Derrida logically argues that this insight throws into crisis the structuralist project of establishing a specifically *human* science. Perhaps this is why the impossible project of grammatology can never be a '*science of man*' (like psychoanalysis or anthropology) insofar as its opening gesture is always to call the '*name of man*' (not to mention that of science) into question (p. 83, Derrida's emphases): the human body, as we will see later on when we turn to the philosopher's reading of the palaeo-anthropologist André Leroi-Gouhran, is *itself* a kind of prosthetic supplement. In the later Derrida's deceptively simple words, there is no natural or originary body or soul: 'technology has not simply added itself, from outside, or after the fact, as a foreign body ... this foreign or dangerous supplement is "originarily" at work and in place of the supposedly ideal interiority of the "body and soul" ' (*Points*, pp. 244–5).

To what extent, though, can Derrida's deconstruction of the opposition between *phusis* and *technē* properly be called a philosophy of 'technology' in the received sense of that term? It is crucial to observe that his interrogation of the opposition between speech and writing, ideality and materiality, the transcendental and the empirical is never to be confused with a simple *valorisation* of what he always calls 'vulgar' technics.[6] As even the most cursory reading of his work reveals all too clearly, the simple gesture of reversing the metaphysical hierarchy in favour of technics would change nothing: 'Nor does it suffice to overturn the hierarchy, or to reverse the direction of the current, to attribute an "essentiality" to technology and to the configuration of its equivalents, in order to change the machinery, the system or the terrain' (*Margins – Of Philosophy*, p. 108). To be absolutely clear from the outset, then, Derrida's deconstruction of the opposition between thought and technology is less an attempt to 'essentialise' a restricted empirical or historical concept of technics than to articulate a new and generalised idea of 'technicity' as one privileged name for the differential spatio-temporal structure of phenomena as such. Such a radicalisation of the classical concept of technicity is one reason why the philosopher remains deeply resistant to allying his work with any reductive materialism (like the Maoist dialecticism of the post-1968 *Tel Quel* group). '[I]f, and in the extent to which, *matter* in this general economy designates radical alterity', Derrida diplomatically responds to the question of the relation between deconstruction and materialism, then 'what I write can be considered "materialist" ',[7] but, as the philosopher is all too aware, this is a very big 'if': materialism in its extant forms – whether historical, biological or scientific positivist – almost always remains complicit with an uncritical empiricism or ontology (see, for example, *Specters of Marx*, p. 170) in his view. For Derrida, in contrast to some of his more unreconstructed materialist champions, it often seems that 'materiality' comes to name – not the mythical empirical real or 'thing in itself' – so much as a kind of algebraic X that marks the spot where absolute alterity interrupts the ipseity of the thinking subject: it is 'a very useful generic name for all that resists appropriation'.[8] Perhaps though, as we will see in the next section, such a 'materialism without matter' also partially explains the later reception of deconstructive philosophy of technicity as harbouring an idealist residuum: what seems to be a critique of idealism from the perspective of a radical material exteriority is charged with always and already idealising its materialist 'outside' in advance. On the one hand, for example, the Deleuze scholar John Protevi accuses Derridaean deconstruction of elevating materiality into a formal, a-historical philosopheme: empirical technology in the Deleuzean sense of an immanent process of machinic becoming is largely disavowed.[9] On the other hand, embodiment theorist Mark B. N. Hansen convicts deconstruction of presiding over a progressive textualisation or 'putting-into-discourse' of matter: empirical technology is granted at best a relative exteriority from the meaning-making acts of the linguistic subject (*Embodying* Technesis, pp. 122–40). In a larger sense, Derrida's philosophy – for all its claim to put into question the name of 'man' – is suspected of still privileging the classic phenomenological milieu of the *ego cogito* as the exclusive site upon which technics impacts: originary technicity – no matter whether it takes the form of the inherently mediated constitution of consciousness or even the radical alterity or blockage beyond which thought cannot travel – remains within the province of an implied thinking subject.

2

In Derrida's later work on technics – *Memoires: For Paul de Man* (1986), *Of Spirit* (1987), *Specters of Marx* (1992), *Archive Fever* (1995) and, particularly, the filmed set of interviews with Bernard Stiegler published as *Echographies of Television* (1998) – he returns continually to the question of the relation between thought and technology. It is still his primary aim to deconstruct a certain metaphysical residue within western philosophy of technology: thinkers as diverse as Marx, Heidegger and Freud are all found guilty of repressing technics as a constitutive force within thought. As we saw in Chapter 4, the later Derrida sees

Heideggerian phenomenology as perhaps the paradigmatic instance of a philosophy of technology that still valorises an originary or essential task of thinking that serves as technology's efficient cause: 'It maintains the possibility of a thought that questions, which is always thought of the essence, protected from any original and essential contamination by technology' (*Of Spirit*, p. 10). For Derrida, in contrast, originary technicity logically forbids any opposition between thought and technology: any allegedly pure or originary 'essence' of technology is always and already contaminated by technics itself. If Derrida always and necessarily begins by positing the *technicity* of thought – as opposed to some more pious *thought* of technology – it is nonetheless possible to ask whether deconstruction's break with Heideggerian phenomenology is as radical as it appears. Such, at least, is the charge of Mark B. N. Hansen who goes so far as to argue that deconstruction even ends up repeating the classic phenomenological gesture of subordinating technology to a primordial and at best quasi-technological ground: Heideggerian Being, crudely speaking, is here replaced by the originary state of empirico-transcendental contamination that is Derridaean *différance* (*Embodying* Technesis, p. 124). Perhaps this charge underestimates Derrida's career-long critique of Heideggerian ontology – whatever else it may be, *différance* is clearly not Being – but what is at least arguable is that *différance* in Derrida performs an analogous *role* to Being in Heidegger: both are pre-phenomenal conditions of possibility for the coming-into-appearance of all phenomena, such as, most obviously, empirical technics. To what extent, then, might even Derrida's theory of originary technicity be guilty of the very charge that he levels against Heidegger: 'Doesn't he suggest that there is a thinking pure of all technics?'[10]

It is certainly true that – for all his criticisms of a certain nostalgic, even ecological remainder within Heidegger's thinking of Being – Derrida's reflections on technics sometimes sound almost self-consciously Heideggerian. As his later, allegedly more ethico-politically oriented, work makes especially clear, originary technicity necessarily involves an interrogation of the truth regime of rational instrumentalism or productionism that Heidegger famously thematises under the figure of *das Gestell*: deconstruction, writes Derrida in deeply Heideggerian language, is a critique of the kind of 'technicist reasoning' that 'installs man in a form that permits him to ensure his mastery on earth and beyond'.[11] To be sure, Derrida never subscribes to Heidegger's grand narrative of western philosophy from Plato to Nietzsche as the progressive reduction of Being to beings, but there is a sense, at least, in which Derrida, too, sees the epoch of contemporary technoscience as one logical fate of a 2500-year-old metaphysics of presence. Perhaps the most powerful example of this logic in the later Derrida is his critique of the metaphysics of 'tele-technologies' – television, email, the internet and so on – in works such as *Echographies of Television* and, in particular, the phenomenon of 'live' (*en directe*) or 'Real Time' (*temps réel*) media. Such Real Time media – whether they take the form of live television or radio programmes, digital streaming of sound and images on the internet or even locative media accessible through wireless networks and handheld interfaces – claim to offer something utterly without parallel in the history of media: the absolute simultaneity of recording and transmission. If Real Time is in one sense absolutely new, though, Derrida also sees it as the logical and historical culmination of the metaphysical project of constructing an absolute spatio-temporal presence (*Echographies of Television*, pp. 5–6 and *passim*): one might go so far as to say that it is the metaphysical spatialisation of time rendered inglorious technological reality. By consistently making such a quasi-ontological move, however, Derrida's philosophy of technology arguably remains faithful to the Heidegger of 'The Question Concerning Technology': both seek to position empirical technics against the backdrop of an immemorial state or condition (call it Being or *différance*) in order to question and understand it. In a larger sense, both also risk reducing technics to the epiphenomenon of a prior and at best quasi-technological condition: whereas the last chapter saw how Heidegger relegates ontic technologies to the vehicle of an ontological revelation, the question here will be whether Derrida subordinates empirical technics to an effect of *différance*.

To Derrida's forensic gaze, of course, what is called 'Real Time' is nothing but a 'mystification': what are called the ' "live" and "real time" are never pure . . . they do not give us intuition or transparency, a perception stripped of interpretation or technical intervention' (p. 5). It is crucial to understand what Derrida understands by time consciousness *itself*, however, if we want to appreciate why Real Time can never really be time and to do this we must go back to his reading of Husserl.[12] According to Husserl's phenomenology of time, of course, what we call the 'now' of perception is in fact a synthesis of the retention of what has just passed (the no longer now) and the protention of what is about to arrive (the not yet now).[13] To prove this, Husserl famously gives a phenomenological account of what happens when we listen to a piece of music. When listening to a melody, he argues, we do not simply hear the individual note that is being played at that exact moment but simultaneously remember the note just passed and anticipate the note to come. Just as we do not hear one note at a time when we listen to a piece of music, he concludes, so we do not perceive time as an endless series of discrete present moments or 'nows'. If Husserl's classic analysis of the human process of temporalisation recognises that what we call the 'present' of perception is always a synthesis of past and future, Derrida famously argues that the phenomenologist does not fully capitalise upon his own insight: Husserl still stresses the originary 'present' impression at the expense of the retentions and protentions that also comprise the process of perception. For Derrida, Husserl's phenomenology of time depends upon an ultimately unsustainable opposition between what the latter calls 'primary retention' (the recollection of the note 'just-passed' that is included in the 'now' of listening to a melody) and 'secondary retention'. This secondary retention is the longer memory we may have of listening to the same melody yesterday, a memory I can hear again in my imagination today, but which (on Husserl's account at least) plays no role in the theatre of perception.[14] In Derrida's view, by contrast, Husserl's decision to include the 'just-passed' in the process of temporalisation is the thin end of the wedge that inevitably opens the door for the entry of secondary memory and imagination as well and, with them, the deconstruction of the residual metaphysics of the latter's phenomenology of time: 'the condition of possibility of the living, absolutely real present is already memory, anticipation, in other words, a play of traces' (*Echographies of Television*, p. 129).

For Derrida, then, it is clear that there is no Real Time *as such* in the sense of a phenomenological time consciousness of the 'now': what seems to be the experience of a living, immediate, if distributed, present in Husserl turns out to be shot through with the absence that is the absolute historical past. It is this aporia of time, more than anything else, which explains why the technological pretension to record and transmit in Real Time must be returned to the finitude of its making: what Real Time seeks to reproduce – the phenomenological experience of a 'now' or living present – never really existed in the first place. Accordingly, Real Time media cannot – no matter how seamless its infinite rolling 'now' of simultaneous recording and transmission may appear – neutralise the non-identity out of which the human experience of time is sutured: 'there is no purely real time because temporalisation itself is structured by a play of retention or of protention and, consequently, of traces' (p. 129). To put it even more strongly, Derrida argues that – far from subsuming *différance* under an absolute 'now' – 'The real time effect is itself *a particular effect of "différance"*' which it can never absolutely reduce (p. 129, emphasis mine). Such is the irreducible difference between what Derrida would call the 'originary technicity' of time – the aporia of difference, spacing and alterity of which the experience of time itself is composed – and the concrete technical inscription of time in 'live' recording. Now, it is precisely this appeal to an 'originary' cause or condition – however artefactual it may be – of which empirical technics seems to be a epiphenomenal effect or manifestation that raises queries about an idealist or transcendental remainder at the heart of deconstruction. If Derrida insists that the aporia of time is the quasi-transcendental condition of (im-)possibility of all technicisation, for example, his former student Bernard Stiegler turns this logic upside down: Stiegler, as we will see in more detail in the next chapter, argues that it is empirical technical inscription *itself* that makes possible all temporalisation – even the aporia of temporalisation that Derrida calls time's originary technicity. Just as Heidegger's belief that ontological time is simply older than the calculable time of technics means that he can presume at least the possibility of obtaining a piously questioning relation to *das Gestell*, so we might argue that Derrida's claim that the aporia of time is independent of, and irreducible to, technicisation also makes possible his own critique of Real Time as mystificatory. In this sense, Derrida's claim that Real Time will never be able to neutralise the aporia of time risks becoming something of a self-fulfilling prophecy: 'the unbridgeable gap between technicisation and temporalisation insures that a critical relation to tele-technologies is in principle always possible'.[15] What would it mean for Derrida's critique of Real Time, though, if there was no such gap between technics and the human experience of time, if the human experience of time were in fact empirically technical through and through?

In the view of critics like Stiegler, then, Derrida's attempt to preserve the originary technicity of time from contamination by empirical technics risks *re-installing* the opposition between thought and technology in the very gesture of deconstructing it. It seems that, once again, empirical technology contains an (albeit aporetic and non-essential) 'essence' that exists entirely independently of it – the aporia of time itself. As we have already hinted in Chapter 3, Derrida's early critique of Freud – for all its radicality – risks *repeating* the de-materialisation of empirical technics at work in psychoanalysis: both see the writing pad as simply a materialised metaphor for an anterior, non-phenomenal condition. For Richard Beardsworth, who has written incisively about the Derrida/Stiegler debate in a review essay of *Echographies of Television*, it is possible to detect a similar blindspot concerning the ontic specificity of technics in the later Derrida. Such 'machines have always been there, they are always there', responds Derrida to Stiegler's claim that Real Time media inaugurate an absolutely new mode of temporalisation, 'even when we wrote by hand, even during so-called live conversation' (*Echographies of Television*, p. 38).[16] By repeatedly returning all technological innovation back to the originary technicity of time itself, Derrida risks repeating the Heideggerian gesture of subordinating ontic technics to a prior ontological ground. Just as Heidegger in the 1940s sees modern industrial production as the outworking of an immemorial *poiēsis* of Being, so Derrida in the 1990s sees postmodern tele-technology as existing purely in order to reveal a prior ('always already') aporetic condition. Perhaps most importantly, Derrida's political solution to the crises of technological modernity in *Echographies* appeals – as we will see in the final section of this chapter – to an absolute 'messianic' future that is every bit as aporetically irreducible to technical calculation as the immemorial past. If deconstruction is 'what happens',[17] though, if Derrida stresses again and again that a politics of technics can never be simply a 'spectatorial critique, a theoretical vigilance' (p. 65), if a great deal of the later Derrida is devoted to arguing that television, email and the internet themselves constitute a 'practical *deconstruction*' (p. 36) of presence that is more powerful and widespread than any philosophical discourse – we can still nevertheless detect a sense in which the deconstruction of technology remains primarily a *philosophical* task:

> And so, how to proceed without denying ourselves these new resources of live television (the videocamera, etc.) while continuing to be critical of their mystifications? And, above all, while continuing to remind people and to *demonstrate* that the 'live' and the 'real time' are never pure, that they do not give us intuition or transparency, a perception stripped of interpretation or technical intervention. Any such demonstration already appeals, in and of itself, to philosophy. (p. 5)

Why this – unexplained and unargued – appeal to the *Deus ex machina* of philosophy? Does it not risk reducing empirical technics back down to its Platonic position as the hypomnetic supplement – however originary or constitutive it supposedly may be – to thought? To what extent might it also risk re-instating something close to the phenomenological *ego cogito* as the thinker of that thought?

What, if I may leave such questions to resonate for a moment, is the upshot of this critique of Derrida's philosophy of technics? It is that the same anthropocentrism we have witnessed in Marx, Freud and Heidegger may still be operative in Derridaean originary technicity: what begins by seeking to devolve the subject onto its environment ends up becoming an object or problem *for* this (supposedly redundant) subject. Accordingly, technics is once again put to work in the economy of subjectification: Derrida, for instance, frequently seems to attribute Real Time media no other role than that of heuristically revealing the always-already technical state of internal time consciousness. Nevertheless, in many ways, such a critique of deconstruction's empirical reduction of technics is itself arguably somewhat reductive, even polemical. To begin with, Derrida would have little truck with what he would undoubtedly see as the unreconstructed empiricism at work in such appeals to the extra-linguistic 'real' of technics, matter, the body which his philosophy allegedly disavows. Such a critique of the philosophy of absolute exteriority – together with an insistence upon the unavoidable *necessity* of the very relative exteriority to the subject that critics like Hansen deem wholly contingent and reductive (*Embodying* Technesis, p. 20) – begins as early as his famous essay on the exteriority of the Levinasian other, 'Violence and Metaphysics' (1964). If Derrida can be charged with transcendentalising *différance*, then (as Geoffrey Bennington has powerfully argued) critics like Beardsworth and Stiegler are equally vulnerable to the charge of seeking to positivise it: what absolutely precedes all phenomena as their condition of possibility is hypostatised – via a gesture roughly analogous to a Kantian transcendental illusion – into a positive object or state of matter.[18] Perhaps a more basic flaw in the case against Derrida is that – in order to make good the claim that he represses or disavows matter – it is forced to simply exclude vast swathes of his more empirically oriented work: a posthumous work like *Paper Machine*, for example, offers exactly the kind of precise reflections on the *specificity* of empirical mnemo-technical apparatuses (including paper, handwriting, typewriting and the computer) that the philosopher's idealism allegedly renders him incapable of producing.[19] In the late text *On Touching – For Jean-Luc Nancy* (2002), as we will now see, Derrida also moves decisively beyond whatever residual phenomenological idealism his work allegedly possesses by proposing a deconstruction – not of the *ego cogito* – but of the body.

3

In his recent book *La Déclosion* (2005), Jean-Luc Nancy has begun to articulate his long-awaited 'Deconstruction of Christianity' (*la Déconstruction du Christianisme*).[20] To Nancy's eyes, the Deconstruction *of* Christianity (double genitive) necessitates an almost unimaginably huge re-writing of the past, present and future of western thought. On the one hand, Nancy shows that the Christian tradition – for all its supposed complicity with metaphysics, ontotheology or logocentrism – is engaged in a process of auto-deconstruction: Christianity deconstructs itself. On the other hand, Nancy also seeks to argue that the post-Christian epoch – from the Enlightenment, through Marx, Nietzsche and Heidegger up to Derrida's own deconstruction – is a logical outworking of that same process of self-deconstruction: what we call deconstruction *itself* is inescapably Christian in origin. If Christianity is nothing other than a perpetual movement of self-overcoming, this means that everything we normally posit as 'beyond' the Christian tradition – the Enlightenment, the Death of God, the so-called 'closure' of metaphysics and even the gesture of deconstruction itself – is inexorably re-absorbed back into it: 'the world that is called modern is itself the becoming of Christianity' (*Le Déclosion*, p. 209). For Nancy, then, the Deconstruction of Christianity poses a critical question to any allegedly secular, post-Christian philosophy. To what extent might we *still* be Christian?

It is this *Déconstruction du Christianisme* that provides Derrida with the occasion to offer a counter-history of deconstruction, of religion and, most importantly for our purposes, of the human body. As Derrida forcefully argues in *Le Toucher – Pour Jean-Luc Nancy*, the problem with the Deconstruction of Christianity is that it risks making something of a Christianity of deconstruction. To start with, Derrida concedes that Christianity can indeed be seen as a certain 'deconstruction' *avant la lettre* even if that gesture obviously passes under other names: Christ's *kenosis*, the mystical *hyperousios* and, particularly, the Lutheran act of *destructio*.[21] For Derrida, we should thus not be surprised to find that what we call 'deconstruction' today retains the memory of Christianity within it: Luther's Protestant *destructio* of Catholic theology in order to return to the originary truth of scripture foreshadows Heidegger's *destruktion* of the history of ontology in the name of a more original, Pre-Socratic experience of Being. However, 'there is deconstruction and deconstruction', as Derrida puts it (*On Touching*, p. 60), and here is the crux of his dispute with Nancy. If Nancy is surely right to say that we cannot simply step 'beyond' Christianity – because the genius of Christianity is that it consists of nothing other than a series of steps beyond itself – Derrida believes that this insight is bought at the cost of a *hyperbolisation* of the Christian tradition: the overcoming of Christianity is always baptised in advance as a Christian self-overcoming. In this sense, Nancy ironically presides over the re-Christianisation of western philosophy when he makes such grandiose claims as: '*All* our thought is Christian through and through' (*La Déclosion*, pp. 208–9).

As Derrida goes on to insist, Nancy's Deconstruction of Christianity is thus a 'difficult, paradoxical, almost impossible task' (*On Touching*, p. 220) – but such impossibility is famously what drives any deconstruction worthy of the name. It is not Derrida's intention to simply rule out Nancy's project, in other words, so much as to put it to the test by trying to imagine what form it must – and must not – take. To begin with, any Deconstruction of Christianity worthy of the name must (like any other deconstruction) inhabit the Christian tradition from the *inside*: we can only begin to overcome Christianity by accepting our utter contamination by it. Nevertheless, we accept tradition only in order to reveal its essential non-identity with itself: what must be affirmed is that there is no single, self-identical or homogeneous 'Christian tradition' in the first place. For Derrida, contra Nancy, the Deconstruction of Christianity is not so much a question of absorbing the entire history of the west into Christianity so much as *turning Christianity itself inside out*: Christianity must be exposed in its very interiority to an outside that is, if possible, even more exterior than the outside it has itself ceaselessly affirmed. If we are to focus upon just one such paradigmatic gesture of deconstruction within the Christian tradition, Derrida argues that it must be the Luthero-Heideggerian gesture of *destructio*: any deconstruction of the Christian tradition, he writes, must begin by 'untying itself from a Christian tradition of *destructio*' whereby the act of self-deconstruction is part of an economy of self-sacrifice, self-purification and thus also self-preservation (p. 60). What, though, might such a Deconstruction of Christianity look like?

To be sure, Derrida touches on many possible answers to this question in *On Touching* – Judaism and Islam to name only the most obvious – but one, perhaps surprising, possibility is originary technicity. It seems, in other words, that a new and radical philosophy of technics fulfils *the very criteria* that Derrida sets out earlier in the book for a new Deconstruction of Christianity. As the philosopher suggests in a remarkable footnote to *On Touching*, technics represents a mode of deconstruction that 'distances itself from the (Luthero-Heideggerian) *destructio*' which is specifically identified as, and with, Christianity (p. 345, n. 26). For Derrida, this new Deconstruction of Christianity takes the form of a labyrinthine reading of the philosophical tradition of hapticity from Greek metaphysics to the modern phenomenology of touch. If it may well seem odd at first to use phenomenology as a means of pursuing a critique of Christianity, Derrida's strategy is clear: Husserl and his successors are – for all the supposed scientific rigour of the *epoché* – by no means exterior to the Christian metaphysical tradition. Perhaps most crucially, the phenomenology of touch – which privileges tactility above all other senses as the most direct and unmediated form of intuition – belongs to a Christian tradition of valorising the body that can be traced all the way back to the Incarnation. In Derrida's version of the Deconstruction of Christianity, however, what is going on is not simply another way of saying that 'everything is always already Christian' but of showing how – in its very 'interiority' – Christianity lies exposed to what lies beyond it.

For Derrida, this critique of the Christian history of the phenomenology of touch – of phenomeno-theology – takes the form of a sustained and remarkable reflection on the place of the hand. What exactly *is* a hand? Who or what does the hand belong to – animal, man, god? Why, in particular, does the history of phenomenology from Husserl, through Heidegger and Merleau-Ponty, up to contemporary figures like Didier Franck and Jean-Louis Chrétien accord such a massive privilege to the hand as the condition of all tactility? In the chapter 'Hand of God, Hand of Man', Derrida offers the following point-by-point answers:

- To start with, Derrida merely notes the absolute indispensability of the hand to the foundational texts of twentieth-century phenomenology. As he shows, Husserl's *Ideas II* are remarkable for the enormous stress they place upon touch: tactility is not merely one sense amongst others but the fundamental condition of our relation both to our own bodies and our life-world. For Husserl, the means of this tactile relation is the hand: what may at first seem to be a mere appendage is in fact the sign of our existence as free, willing, spontaneous, affecting and auto-affecting egos (*On Touching*, p. 160). Perhaps more generally, the hand is also the index of a metaphysics of immediate, intuitive presence whether it be of our own bodies or of the world: the finger that touches something is simultaneously and indivisibly touched *by* that something in a way that applies for no other sense. (We will have something to say about the obvious metaphysics of this idea of the touch later on.) In twentieth-century phenomenology, of course, such a valorisation of the hand is remarkably persistent – beginning of course with Heidegger's own paradigmatic attempt to articulate *Dasein*'s engagement with the world through the concept of the 'ready-to-hand' (*Zuhandenheit*) (see Derrida's 'Le main de Heidegger (*Geschlecht* II)', pp. 415–71 for an elaboration of this argument).

- However, what none of this tells us is *why* the hand is so indispensable to the history of phenomenology. It may well be that tactility is the most direct and immediate form of intuition – though Derrida will locate problems here too – but none of this explains why the hand alone is deemed the privileged or exclusive vehicle of touch. After all, it is by no means the only part of the human body that is capable of touch. To pursue this argument even further, we could ask why the philosophy of haptology chooses to focus on the *human* body at all as opposed to say the animal – are human beings really the only beings that are capable of touch, and thus a relation to their own bodies and to the world? Now, Husserl does not quite say this and there are good phenomenological grounds for focusing an analysis upon human beings rather than, say, animals: we should always begin with the being that is closest to us and nothing is closer than the being that we ourselves are (*On Touching*, pp. 166–7). Even so, a rigorous phenomenology should suspend all reference to the real nature of both the perceiver and the perceived – the 'who' and the 'what' – and concentrate on the phenomenon alone, and there are grounds for wondering whether this is really the case here. Just as Husserl's phenomenology of the sign fails to bracket off a certain metaphysics of the voice, so his phenomenology of touch is contaminated by a metaphysical anthropology. For Derrida, that is, what is at stake in the phenomenology of the hand is a residual anthropocentrism that establishes a rigid hierarchy and teleology: the only being that can genuinely touch – and thus constitute himself as free, spontaneous agent – is man (p. 154).
- Perhaps we can go still further. As Derrida goes on to argue, what we are dealing with here is not simply an anthropology but a *theology*: the phenomenology of touch – continuous, immediate and indivisible – inevitably touches upon a theology of touch, of contact, of consubstantiality of being that finds its paradigmatic example in the doctrine of the Incarnation. To prove this – obviously somewhat speculative – point, Derrida turns to Jean-Louis Chrétien's phenomenology of the 'hand of god'.[22] On one level, of course, nothing could be further from the scientific rigour of Husserl than this kind of phenomenological theology: it is no coincidence that Chrétien was, along with Jean-Luc Marion and Michel Henry, the principal target of Dominique Janicaud's polemic against the '*tournant theologique*' in French phenomenology.[23] Nevertheless, we can draw a direct line from the metaphysics of the touch in Husserl's *Ideas II* to the theology of the touch in Chrétien's *L'appel et la réponse*. For Chrétien, it is only when we are touched by the *divine* that we achieve

> the full, immediate and intuitive plenitude promised by Husserl: only the hand of god genuinely delivers up the spatial and temporal presence offered by the hand of man. If we commonly think of the 'hand of god' as nothing more than a metaphor or analogy drawn from the sensible, finite hand of man, in other words, Chrétien turns this logic on its head: it is the hand of man that is a figure for the real, authentic and immediate touch offered by the hand of god. Now, this is not to say that Chrétien simply opposes the philosophical privilege accorded to human touch above all others because what defines the spiritual touch for him is precisely the doctrine of the Incarnation, that is to say, the becoming-human of the spiritual. Rather, we might say that the hand of god is paradoxically the only means of ensuring the historical privilege accorded to the hand of man: it is the hand of god that *gives* the hand of man, that *reveals itself to* that hand and *redeems* it from its creaturely finitude. Such is how the philosophical priority of tactility is secured by theology: the human hand is called by, and responds to, the spiritual touch of the Incarnation.

In this exemplary deconstruction of the philosophy of hapticity, Derrida even tempts us to wonder whether Chrétien's analysis of the 'hand of god' is less proof positive of a 'theological turn' in phenomenology (à la Janicaud) than the final outworking of the anthropotheological assumptions that have *always* been in play in discussions of the human hand: Christianity deconstructs itself in order to re-appear as the phenomeno-theology of touch.

What form, though, might a deconstruction of such a Christian phenomenology of touch take? It is, once again, a question neither of opposing such a history from some spuriously external position, nor of hyperbolising it, but rather of exposing its own 'outside' within it: the anthropotheological hand is not at one with itself. As Derrida tantalisingly suggests in one of those passing asides in his work that always contain philosophical depths, it is possible to offer another history of the hand that exists independently of all anthropotheology: the 'history of the hand' remains 'impossible to dissociate' from the 'history of technics', he writes (p. 154). In 'Hand of Man, Hand of God', Derrida does not particularly elaborate upon this remark but, put in the context of his earlier work on technicity, the trajectory of his argument can again be elaborated step by step:

- To Derrida's way of thinking, an originary technicity – mediation, spacing or substitutability – always exists at the heart of the phenomenology of touch: what appears to be the immediacy, continuity and indivisibility of the touch is always mediated by and through alterity. It is thus clear that in this late text Derrida is turning full circle: what *Speech and Phenomena* did for the metaphysics of speech, *On Touching* now does for contact. As the philosopher works through Husserl, Merleau-Ponty, all the way up to Nancy's own discussions of the plasticity of the body in *Corpus*, he finds that mediation always insinuates itself between the apparent spatio-temporal immediacy of contact: there is a spatio-temporal difference at work even between the simultaneity of the touching and the touched. For Derrida, in the context of a discussion of Franck, it even becomes possible to ask whether the 'phenomenology of touch' is, strictly speaking, an oxymoron: what makes it possible for us to analyse touch as a *phenomenon* – the reduction or bracketing of the world in all its contingency, variability and changeability – is precisely what forbids any *real* experience of contact itself. The phenomenon of contact – with all the immediacy, simultaneity and presence it entails and guarantees – is only constituted on the basis of the suspension of real contact. In this essential mediation that constitutes our very experience of the 'immediate', the originary technicity of perception, affection and auto-affection is born: 'what calls for "technics", then, is the phenomenological necessity itself' (*On Touching*, p. 230).

- Perhaps we can best witness this technicity in operation by returning to the question that drives Derrida's entire discussion: what exactly *is* a 'hand'? Who does it belong to? Where does it come from? To be sure, anthropotheology provides one answer to that question: the hand is the hand *of* man – the hand that defines man as man and not animal, the hand that enables him to touch himself and his world, the hand that guarantees his privileged status as free, spontaneous self-moving, affecting and auto-affecting ego. However, we now know that there is another history of the hand, and another Deconstruction of Christianity, albeit one that is only alluded to by Derrida himself in the footnotes to the text (p. 345, n. 26; p. 362, n. 34), but one that I would like to tease out here. For André Leroi-Gouhran – whose groundbreaking work *Le geste et la parole* is discussed by Derrida as early as the *Grammatology* and referred to once again in the notes to *On Touching* – the hand is the engine that drives the process of hominisation itself: everything begins with the hands and the feet.[24] By obtaining an erect posture, prehominid man liberated the hands for tool-use and tool-use, in turn, liberated the lower jaw for language and enabled the cerebral cortex to fan out across the brain pan. In Leroi-Gouhran's account, as we have already seen in Chapter 1 of this book, the story of humanity is thus really the story of a progressive locomotive self-liberation or 'exteriorisation' which begins with the secretion of the primate's skeleton, nails and teeth, continues through the fabrication of the flint axe, and carries on all the way up to the computer revolution: 'The whole of our evolution has been oriented towards placing outside ourselves what in the rest of the animal world is achieved inside by species adaptation' (*Gesture and Speech*, p. 42).

- So, what, then, *is* a hand? In Derrida's view, a radicalised re-reading of Leroi-Gouhran's palaeo-anthropology enables us to supply a counter-history of hapticity to that supplied by the history of anthropotheology. It is Leroi-Gouhran's position, recall, that hapticity is one of the first instances of human evolutionary exteriorisation. However, Derrida seeks to divest this process of primordial exteriorisation of any concept of an anthropological origin: 'exteriorisation', he writes, has 'always already begun' (*Of Grammatology*, p. 85). For Derrida, to recall his famous logic of the supplement, human essence, being or substance does not pre-exist exteriorisation but is rather constituted retroactively on the basis of this becoming-exterior. Such, then, might be the basis of a deconstructive haptology: what we call the humble human hand is an *originary supplement* – the first tool or prosthesis – out of which the being of the human itself is constructed. If Derrida is by no means the first philosopher to speak of the hand in prosthetic terms – Aristotle calls it the tool of tools in *De Partibus Animalium* whereas Descartes famously compares the entire body to a tool or machine in the *Meditations on First Philosophy* – it is important to stress here that the former's position remains radically different from that of metaphysical haptology: the hand is not a prosthesis of a pre-existing consciousness – which thus acts as its efficient cause – but rather consciousness is the retrospectively internalised after-effect of the hand. Perhaps more pertinently, such a technical counter-history of the hand also calls into question the anthropological privilege accorded to the human by Christianity and, latterly, the phenomenology of touch: what phenomenology calls the hand of *man* – the hand as the sign, attribute and extension of a pre-existing human being who occupies pole position in a teleological hierarchy – can no longer be called an attribute (essential or accidental) of man. In a more profound sense, what appears to be most proper to the human – to constitute our privileged status within the world as free, spontaneous (self-) affecting egos – turns out to be that which

> utterly ex-propriates the human: our immediate experience of ourselves is mediated all the way down because the very agency of that self-experience – the hand – is *itself* an originary medium.

For Derrida, such a deconstruction of the body is also and at the same stroke a deconstruction of Christianity: a body that is 'originally and essentially open to... *techne*', he writes, is also one that is involved in a '"deconstruction of Christianity", of the Christian body... of Christian "flesh"' (*On Touching*, pp. 218–19). It is his hypothesis, remember, that the phenomenology of touch is inseparable from a Christian theology of immediacy, plenitude, consubstantiality and ultimately Incarnation. We would have no phenomenon of touch – no concept of the spatio-temporal identity between the touching and the touched – without Christianity. If Christianity tells us that the hand of man always contains within it the trace of a real, divine presence, though, Derrida effectively turns that hand inside out, like the fingers of a glove, to expose the exteriority from which it is stitched: what emerges is something irreducible to the Christian tradition which is nonetheless at the very heart of that tradition. Such, perhaps, would be a Deconstruction of Christianity that did not make a Christianity of Deconstruction.

In my view, the deconstructive theory of embodiment set out in *On Touching* goes a long way to meet the charge that Derrida's philosophy of technics harbours an idealist phenomenological remainder. It is not simply that Derrida accepts the banal truth that 'thought has a body' but that such a body remains absolutely resistant to phenomenological reduction: the hand can never become present *to* consciousness because it is itself the historico-empirical condition of possibility *of* consciousness. At the same time, this body – far from occupying a pre-technical organic state – is always open to, and formed by, its relation to technics: the history of the hand is, as Derrida himself writes, impossible to dissociate from the history of technics. Yet it would be naive to think that *On Touching* will simply put an end to suspicions of a residual allegiance to idealism within Derrida's theory of originary technicity. To start with, Mark Hansen and other embodiment theorists would undoubtedly suspect that what Derrida calls 'the hand' is – for all its apparent fleshly recalcitrance to phenomenological reduction – still implicitly modelled upon a prior linguistico-phenomenological analytic of the aporia of origin. Such a claim is not a rehearsal of the old cliché that deconstruction simply reduces the world to a text, but rather that it textualises the world: what is nominally the other of thought remains thought's *own* other. By the same token, it could be argued that Derrida's theory of embodiment still de facto (if not de jure) preserves the very anthropotheological privilege accorded to the hand by the history of philosophy from Aristotle to Husserl. Just as *Echographies of Television* focuses on what new media means *for* a metaphysical phenomenology of time, so *On Touching* holds tight to technicity's implications for a metaphysical phenomenology of space. Both end up perpetuating the reign of the *ego cogito* – capable of intentionalising a moment of spatio-temporal presence – precisely in order to critically unmask it. Perhaps, though, one final question we might address to Derrida's theory of technics (and one that goes largely unasked by his techno-materialist critics) is whether it is guilty of *the very charge* it levels against Nancy's Deconstruction of Christianity, namely, of promulgating a quasi-transcendental – indeed quasi-religious – philosophy of history at the expense of a quasi-empirical genealogy. If there is an 'elephant in the room' of Derrida's critique of Nancy's re-Christianisation of philosophy – which is as conspicuously absent from *On Touching* as it is omnipresent everywhere else in the later work – it is Derrida's own so-called 'religious turn', which (just as controversially as Nancy's) deems it historically and strategically necessary to think deconstruction through the terms provided by the Judaeo-Christian tradition: 'I am obliged ... for essential reasons of time and of place, of culture', the philosopher writes in the essay 'Faith and Knowledge', 'to continue giving it [deconstruction] names marked by the Abrahamic tradition'.[25] To what extent, then, is Derrida's turn to religion also a turn *away* from technics?

4

In Derrida's later work, he famously advances a new, startling and controversial definition of his philosophy: deconstruction, he writes, is 'a thinking that "repeats" the possibility of religion without religion'.[26] It is possible to witness this non-identical repetition of the religious throughout his post-1980 work as Derrida turns again and again to Judaeo-Christian themes, problems and traditions in his thought: apophasis, givenness, confession, sacrifice, hospitality, faith and knowledge, the sacred, the impossible and, perhaps most importantly, the messianic. As the philosopher famously adumbrates in *Specters of Marx*, for instance, deconstruction can even be called 'a messianism without messianism' insofar as it affirms 'the "yes" to the *arrivante(e)*, the "come" to the future that cannot be anticipated' (*Specters of Marx*, p. 168). To be sure, Derrida's religious 'turn' remains one of the most hotly contested areas of his thought and critics of his work have already narrated its significance in diametrically opposed ways. On the one hand, for instance, Kevin Hart, John D. Caputo and Hent de Vries have affirmed the religious or theological turn as the logical extension – perhaps even the perfection – of his articulation of a radical alterity or openness at the heart of thought.[27] On the other hand, however, Richard Beardsworth, Bernard Stiegler and most recently Martin Hägglund have attacked the religious turn as an illicit transcendentalisation or theologisation of this excess or contingency within thought.[28] If the former group of critics all endorse, more or less, the view that deconstruction is a 'religion without religion' – a purely formal structure of waiting, anticipation or expectation as opposed to a theological or political dogma (*Specters of Marx*, pp. 167–8) – the latter generally attack such a reading as, once again, a disavowal of a certain non-reductive empiricity or materiality *within* alterity. For Beardsworth – who frames this dissensus most dramatically in the conclusion to his *Derrida and the Political* (1996) – it even becomes possible to speak of two (seemingly incommensurable) futures for Derrida's thought in the wake of the 'religious' turn: either we subscribe to a 'left-wing' deconstruction – which takes its lead from the earlier quasi-empirical analyses of writing, technics and matter – or a 'right-wing' deconstruction – which emerges out of the later quasi-transcendental or even theological readings of the promise, negative theology and literature (*Derrida and the Political*, pp. 156–7).[29] To what extent, then, are we compelled to choose – even though the later Derrida himself frequently and plausibly insists that we *cannot* choose ('Faith and Knowledge', p. 28) – between a 'religious' and a 'technical' deconstruction?

For me, we can witness what is at stake in the relation between deconstruction, religion and technics most clearly in two important essays: Derrida's own 'Faith and Knowledge: The Two Sources of "Religion" Within the Limits of Reason Alone' (1994) and Stiegler's reply to Derrida 'Fidelity at the Limits of Deconstruction and the Prosthesis of Faith' (2002). It may seem at first glance from reading these texts that Derrida and Stiegler come to exactly the same conclusion about religious faith and technics: each is inextricably folded into the other. As will become clear, though, this apparent agreement conceals a larger and more important disagreement that goes to the heart of Derrida's philosophy of technics. What form does this differend take?

- In 'Faith and Knowledge', Derrida famously argues that religion (which includes everything from faith and revelation at one extreme to philosophical theology at the other) and reason (which again is defined broadly enough to include the categories of critique, science and technological rationality) share the same aporia of origin ('Faith and Knowledge', p. 28). Both religion and reason, he contends in a radicalisation of Heidegger's concept of the *Zusage* or pledge, presuppose an immemorial 'performative of promising' (p. 18) that exceeds any determinate speech act, intentional agency, empirical event or transcendental condition. If all speech acts share the same radical faith, then this common origin folds religion and reason into a logic of mutual contamination from which no living present escapes (pp. 63–5). Such an 'elementary act of faith' (p. 18) – the basic and irreducible openness to the future *qua* future – is the precondition of any address to the other whatsoever whether religious or rational.

- Yet, it is precisely this pre-originary faith that binds religion and technics together which is the subject of Stiegler's text, 'Fidelity at the Limits of Deconstruction and the Prosthesis of Faith'. This essay, which at first glance looks like a simple commentary upon 'Faith and Knowledge', actually offers a sophisticated contrapuntal reading of what it sees as a fatal 'hesitation' concerning technics in Derrida's own text ('Fidelity at the Limits of Deconstruction', p. 241). It is Stiegler's hypothesis (which, as we will see in Chapter 5, is informed by his own distinctive philosophy of technics) that Derrida's 'quasi-transcendental' version of deconstruction increasingly hardens into a *simply* transcendental philosophy that occludes the very technicity it claims to embrace. To begin with, Derrida's insistence upon the absolute irreducibility of *différance* to 'any sensible plenitude, audible or visible, phonic or graphic' (*Of Grammatology*, p. 62) constitutes what Stiegler calls a 'somewhat surprising declaration of *independence*' from the phenomenal world ('Fidelity at the Limits of Deconstruction', p. 253). If *différance* clearly cannot be reduced to some historical event or empirical object in the world, Stiegler nonetheless quite rightly insists that it cannot exist in some Platonic 'other world' either unless it really is a transcendental subject, cause or condition after all. Perhaps another way of making this same point is to say that *différance* must *necessarily* (and not merely contingently) fall into the empirical world: this structural fate is what puts the 'quasi-' into its 'quasi-transcendental' status. In Stiegler's words, what Derrida calls the aporia of origin must at the very least pass '*through*' empiricity even if it is ultimately irreducible to it (p. 260).
- For Stiegler, then, we must supplement Derrida's quasi-transcendental philosophy of the aporia with a quasi-empirical genealogy of its passage *through* the empirical world and the vehicle of that passage is,

> of course, technics: 'the inverting logic of the supplement outlined here requires a history of the supplement, *one which is also a history of technics [les techniques]*' (p. 248). It is not enough to argue that religion and technics share an immemorial origin that is older than any history, as Derrida does, because such an appeal to an ultra-transcendental origin risks recapitulating the very opposition it seeks to deconstruct. As we saw in the case of Nancy, the deconstruction of religion ultimately ends up being carried out by something (however formal, contentless and pre-religious it claims to be) very close to religion itself – a radical 'faith' no less. Such a quasi-transcendental 'always already there' must, Stiegler argues, be supplemented by a quasi-empirical 'but *through* here' if it to avoid this religious fate: what Derrida sees as the *absolute* past of the aporia of origin must necessarily be mediated through what Stiegler calls the *relative* past of history (pp. 254–5). In this account, every faith in a pre-phenomenal, quasi-transcendental point of origin – whether we call it 'God' or '*différance*' makes no difference in this context – must always bear witness to the determined empirical history – whether it be holy scripture or alphabetic writing – in which that immemorial past finds itself necessarily yet imperfectly actualised: 'Quasi-transcendental. But *through* empiricity. Empiricity invested with spirit, with the spirit, for example, of the *letter*: the empirical mark of the trace common to geometry, the Bible, the New Testament and the Koran' (p. 260).

In the spirit and the letter of Stiegler's own text, then, Derrida's account of the radical faith that underwrites the determined opposition between religion and technics must itself be underwritten with a radical and empirical technicity: '*Technics is the condition as much of science and knowledge as of religious faith*' (p. 259, Stiegler's emphasis).

To what extent can we resolve the differend between Derrida and Stiegler, between faith and technics and, by extension, between the two very different fates for deconstruction they seem to map out? It is possible to detect two main critical lines of response to this question which generally seek either (1) to endorse Stiegler's critique of an excessive formalism or transcendentalism in Derrida (Beardsworth, Hansen) or (2) to question an (accidental or structural) positivism within Stiegler's own critique (Bennington, Wills). As the battle lines have become drawn, a certain polemicism has also perhaps inevitably crept into the debate with the key question increasingly becoming whether Derrida or Stiegler's position is ultimately 'right'. Yet, taking a step back, one might wonder whether – like Derrida's famous debate with Searle over speech act theory – the Derrida/Stiegler exchange ever actually takes place. To be sure, Derrida and Stiegler both promulgate something they call 'originary technicity', but it is clear that the two thinkers mean something very different *by* 'technics', not to mention the sense in which it is originary. On Derrida's side, technics is the quasi-transcendental aporia of spatio-temporal difference, repetition and alterity to which all empirical phenomena are originarily subject: 'the technical is the possibility of faith, indeed its very chance' ('Faith and Knowledge', p. 47). On Stiegler's side, however, technics is the empirical inscription onto mnemo-technical supplements to which that quasi-transcendental aporia is itself necessarily behoven: the aporia of origin 'depends' on technics ('Fidelity at the Limits of Deconstruction', p. 253), he insists, technicity is 'essential' to the aporia (p. 254). Perhaps – for all the controversy the Derrida/Stiegler exchange has generated – one might be tempted to conclude that *both* positions are legitimate insofar as they describe the same empirico-transcendental state of contamination albeit from different ends of the spectrum: the aporia of origin is neither inside the empirical (Derrida) nor outside it in some transcendental other world (Stiegler). Such a conclusion would arguably do more justice to the full complexity of the two philosophers' positions than the recent controversy permits: Derrida always goes to great pains to distinguish between *différance* and classical concepts of transcendence in his early work ('I am not invoking an *other world*, an alibi or an evasive transcendence' (*Writing and Difference*, p. 57)) whereas Stiegler never seeks to reduce *différance* to a simple technological *positum* ('Fidelity at the Limits of Deconstruction', p. 263 and *passim*). If Geoffrey Bennington seeks to charge Stiegler with confusing a 'valid and necessary claim about originary technicity' with 'a positivistic claim about techno-science' ('Emergencies', pp. 195–6), for example, Stiegler might reasonably reply that this is less a simple confusion than a subtle, deliberate and exhaustively worked out argument: what he is actually claiming is that we would have no access to Derrida's quasi-transcendental logic of originary technicity *without* its inscription in empirical technics, not that they are exactly the same thing ('Fidelity at the Limits of Deconstruction', pp. 254–5). Why – if there is no relation between originary technicity and empirical technics whatsoever – continue to call the former a species of 'technics' at all? Where exactly does originary technicity exist if it is entirely independent of the empirical world and yet not a transcendental concept either? To what extent could we even think it if it is neither an empirical state nor a transcendental condition?

5

What, then, can we conclude about Derrida's own theory of originary technicity? It is beyond dispute that Derrida's thought represents the most powerful philosophy of technology we have confronted so far: the deconstruction of western metaphysics is always the deconstruction of the sovereignty of *phusis* over *technē* (*Margins – Of Philosophy*, p. 17). As we have seen over the course of this book, Derrida offers not only a forensic diagnosis of what we have seen to be the persistent repression or exclusion of technics within the western philosophical tradition from Plato to Heidegger but also the most far-reaching exploration of what the inclusion of *technē* might involve for everything from the human to the divine. However, all that being said, we might also argue that there is a sense in which even Derrida's theory of technics describes the philosophical circle that he so famously articulates in his classic critique of the human sciences, 'The Ends of Man': '[w]hat is difficult to think today is an end of man', Derrida writes, which would not be 'a teleology in the first person plural' (*Margins – Of Philosophy*, p. 121). For Derrida, originary technicity may not be 'an attribute – essential or accidental – *of* man' (*Of Grammatology*, p. 244) but, nonetheless, there are moments where even deconstruction cannot help but turn this prosthesis of origin into a question *for* 'man': technicity effectively becomes a means of describing what Christopher Johnson calls the enduring, indeed machinic, trope of 'what-it-means-to-be-human'.[30] If the deconstructive logic of the supplement apparently breaks with every classical concept of the prosthesis, in other words, it often still remains a means of revealing *to* what-is-called-'man', the always-already-technicised status of its consciousness, of its embodiment, of its experience of time, space and of presence as such. Such a deconstruction of man arguably remains a virtual teleology in the first person plural – not because it furnishes any new definition of what is 'proper' to our human being – but at least insofar as it retains a philosophical investment in the propriety of the *question* 'But who, we?' (*Margins – Of Philosophy*, p. 126). Perhaps, in this sense, Derrida's philosophy of technics represents the most attenuated version we have seen thus far of the anthropological machine which grants the human subject a spectral afterlife from which it can – impossibly – bear witness to its own deconstruction. In the next chapter, we turn to an important contemporary attempt to correct what it sees as this idealist remainder within Derrida's deconstruction: Bernard Stiegler's philosophy of technology.

6
Time

> The whole question is whether any ... evaluative hierarchy – according to which technics remains only *on one side* (of an opposition), itself not *constitutive* of individuation – in fact remains 'metaphysical'. (*The Fault of Epimetheus*, p. 14)

In the contemporary philosophical moment, the theory of originary technicity is associated with one name above all others: Bernard Stiegler. It is for his (apparently still incomplete) magnum opus *La Technique et le temps* (*Technics and Time*) (1994–) that Stiegler remains best-known in the Anglophone world.[1] As the self-consciously Heideggerian title of this work already implies, Stiegler's philosophy of technology is founded on the audacious claim that there is an essential relation between technics and time: what phenomenology calls the human capacity to 'temporalise' – to organise its experience of the flow of time – is constituted through, rather than merely supplemented by, mnemo-technical prostheses (*The Fault of Epimetheus*, p. 17). Yet, if our experience of time is originally technical, it seems that this condition has remained largely occluded in the western metaphysical tradition, as it evolves from Plato to Heidegger. For Stiegler, western philosophy from Greek metaphysics to modern phenomenology instead institutes a series of oppositions *between* technics and time that relegates the former to a purely incidental or supplemental position.

To Stiegler's way of thinking, by contrast, we can no longer oppose technics and time: *all* time – whether anamnestic or hypomnetic, qualitative or quantitative, calculable or incalculable, messianic or historical – is technical all the way down. Such, at least, is the major thesis of *La Technique et le temps* and it is one that Stiegler pursues through a (complex, uneasy and still arguably misunderstood) conceptual and methodological re-organisation of the empirico-transcendental distinction. On the one hand, he pursues a quasi-transcendental critique of the repression or forgetting of technics in western philosophy of time from Plato up to and including Heidegger and his intellectual mentor Derrida. On the other hand, he proposes a quasi-empirical genealogy (drawing on such figures as Leroi-Gouhran, Simondon and Valera) of the evolving historical and political relation between technical prostheses and human processes of temporalisation from the birth of alphabetic writing to the epoch of digitalisation.

For Stiegler, as we will see in detail throughout this chapter, contemporary capitalism produces what he calls the 'hyper-industrialisation' of memory, temporal objects and consciousness: the human capacity to temporalise – itself inherently technical – is now subject to a techno-economic act of appropriation that is without parallel in history. Perhaps the most prominent – and yet for crucial reasons still largely invisible – example of this process is the programming of so-called 'live' or 'Real Time' phenomena by the globalised film and television industry: Real Time, in his account, collapses the complex human/technical temporalisations of time into an endless synthetic present time that can only be consumed passively and uncritically. If Stiegler's philosophy arguably represents the most radical and ambitious attempt to articulate a state of originary technicity, it has nonetheless received a somewhat uneven reception within the contemporary intellectual scene. In the years since the first volume appeared, *La Technique et le temps*' re-working of the empirico-transcendental opposition has been routinely accused of promulgating everything from a residual phenomenological idealism to a crude technological determinism.

This chapter seeks to explore Stiegler's philosophy of technics from its account of the birth of the human species millions of years ago to the crises of individuation brought about by the industrialisation of time, memory and desire under contemporary capitalism. To what extent does Stiegler's version of originary technicity succeed in deconstructing the enduring historic opposition between technics and time, or – like its predecessors – does it, too, fall into the trap of valorising a pure, technics-free temporality?

1

In 'Prometheus's Liver', a chapter in *La Technique et le temps 1: La Faut d'Épiméthée* (*The Fault of Epimetheus*) (1994), Stiegler offers a now-famous reading of the myth of Prometheus (pp. 185–203) which also serves as a useful propadeutic to his project as a whole. To be sure, everyone knows the story of Prometheus – the Titan who steals fire from the Gods and is eternally punished for his crime – but the role played by his idiotic brother Epimetheus in this narrative has been largely forgotten even though it scarcely makes sense without him. According to Plato's account in the *Protagoras*, the two brothers are ordered by the Gods to equip each mortal species with different and complementary powers or qualities (*dynameis*) to enable them to survive. However, Epimetheus rashly persuades Prometheus to let him do the job himself, and sets about endowing some animals with speed, others with strength and so on (*'Protagoras' and 'Meno'*, 320c–2a). Now, in his haste Epimetheus forgets to allocate any powers at all to human beings – leaving them entirely naked and defenceless – and this is why Prometheus is forced to steal the gift of skill in the arts (*technai*) from Hephaestus and Athena by way of compensation for this loss. For Stiegler, this mythopoeic account of the origin of man contains a crucial insight into the status of the human that will form the basis for his own philosophy: humanity is constituted by an originary *lack* of defining qualities allegorised by the Eipmethean error. Such is what the philosopher calls our necessary 'default' of origin (*le défaut qu'il faut d'origine*). What, then, comes to fill this originary lack?

It is actually possible to give a one word answer to this question: epiphylogenesis. According to Stiegler, we can distinguish between three different layers of memory: (1) genetic memory (which is the biological memory that is programmed into our DNA), (2) epigenetic memory (which refers to the experience we acquire during our lifetime and is stored in the central nervous system) and, finally (3) epiphylogenetic memory. To offer a quick definition, epiphylogenetic or tertiary memory is the mnemo-technical process of retention where successive genetic and epigenetic memories come to be deposited in technical systems or artefacts like tools, cave paintings, archives and so on. For Stiegler, it is significant that human beings are the *only* living beings who possess the third form of memory: we – apparently alone amongst all life forms on earth – live by recording, stockpiling and transmitting our experiences to others in the form of technical artefacts and support systems (*The Fault of Epimetheus*, pp. 176–7). Such is what the Epimethean 'man without qualities' is granted by Prometheus in lieu of his *dynamis*: technics. Perhaps the closest philosophical predecessor to Stiegler's epiphylogenetic account of the human – where we exist by putting our experiences outside ourselves – is André Leroi-Gouhran's palaeontological theory of human history as a process of exteriorisation (*Gesture and*

Speech, p. 42). If Leroi-Gouhran was the first to articulate the becoming-exterior of the human, however, we saw in Chapter 5 how Derrida's logic of an originary supplement that precedes and underwrites the opposition between the human and the technical sought to divest this process of any spiritual, biological or anthropological essence: hominisation after Leroi-Gouhran must be seen as a process of 'exteriorisation' that has 'always already begun' (*Of Grammatology*, p. 85). In the previous chapter, we also began to see that Stiegler has a subtly different philosophical itinerary from Derrida – the former seeks to construct a material genealogy of specific, determinate supplements from the flint tool to digital technologies, crudely speaking, whereas the latter attempts to articulate a quasi-transcendental logic of the supplement – but in many ways both arrive at the same deconstruction of the *anthropos:* what we call the human is a process of exteriorisation that paradoxically has no antecedent or pre-existing interiority (*The Fault of Epimetheus*, p. 141).

Yet, if epiphylogenesis has important implications for our understanding of human being, it also transforms our understanding of technology. It is once again a question of demonstrating how, as we have seen throughout this book from Marx to Derrida, the human is constituted *by* technics as much as it constitutes technics: '[t]he prosthesis is not a mere extension of the human body; it is the constitution of this body *qua* "human" ' (pp. 152–3). As *The Fault of Epimetheus* goes on to show, when we still deludedly speak of human 'interiority' – whether it be a soul, free will, consciousness or some other property that is deemed to pre-exist all exteriority – we are putting the human cart before the technological horse: any interiority has actually been *constituted retroactively* by the process of technological exteriorisation (p. 142). For Stiegler, the theory of epiphylogenesis thus calls for an entirely new account of the relation between the human and the tool – inside and outside – as two elements in the *same* dynamic, recursive feedback system: 'man (the interior) is essentially defined by the tool (the exterior)' (p. 142). By transforming his environment through tools, the human finds himself and his own nature transformed because, as we have already learnt from Marx's anthropology, he has no nature outside that environment. Now, what is at stake here no longer seems to be an *opposition* between interiority and exteriority – terms which increasingly cease to have any meaning as they collapse into one another – but rather an 'originary complex' in which the two terms 'far from being opposed, compose with one another' (p. 152). Perhaps the most remarkable example of this originary complex (developed through a close reading of Leroi-Gouhran's palaeontological theory) is the co-evolution of the flint tool and the cerebral cortex in Zinjanthropic man to the point where it becomes impossible to say which 'caused' which: 'the cortex is determined by the tool' as much as 'the tool by the cortex' (p. 176). If it is possible to detect a number of implicit and explicit scientific antecedents to Stiegler's theory of epiphylogenesis – from Lamarckian evolutionary biology to Valera's idea of a structural coupling between organism and environment – an enduring if underrated influence is Gilbert Simondon's theory of the pre-individuated metastable field out of which individuated entities emerge. In Stiegler's account, the originary complex between the 'who' of humanity and the 'what' of technics is 'transductive' in Simondon's technical sense of a relation whose constitutive terms do not pre-exist the pre-individuated field or environment from which they are formed (*The Fault of Epimetheus*, p. 152; *Désorientation*, pp. 14–15).[2]

To be sure, Stiegler pursues the logic of this transductive relation between the 'who' and the 'what' in many different directions across the different volumes of *La Technique et le temps* – both philosophical and empirical – but his most far-reaching thesis remains the same: there is an essential relation between technics and the human experience of time. It is not enough to attempt to think technics *in* time (in the sense of technology's own evolution through history) nor as a question *of* time (in the sense that using even the most basic tools inevitably involves a sense of anticipation and retention on the part of the operator) but something more radical: we must understand technics *as* time in the sense that technical objects are what make possible our phenomenological temporalisation of time into past, present and future in the first place (*The Fault of Epimetheus*, pp. 27, 83). As Stiegler goes on to argue through another radicalisation of Leroi-Gouhran, the process of exteriorisation – of putting ourselves outside ourselves – creates the horizon of future possibilities against which the process of temporalisation becomes possible: '[t]here is no anticipation, no time outside of this passage outside, of this putting-outside-of-oneself and of this alienation of the human and its memory that "exteriorisation" is' (p. 152). For Stiegler, Zinjanthropus' simple gesture of striking a stone at 90 degrees in order to produce a sharp-edged tool – an act that takes place long before the evolution of so-called 'symbolic' consciousness – sparks off an ever-complexifying process of anticipation and recollection (p. 152). Just as there is no interiority before exteriority, so there is no temporalisation before the tool: the human invents the tool which, in turn, invents him (pp. 153–4). Perhaps the most important way in which the technical constitutes time for Stiegler however – and here once again we encounter the extent to which, for all his critique of Heidegger, he remains deeply Heideggerian – is that it embodies the problematic of human *finitude*: we exist in a world that is not of our making and which will continue to exist after our death. If what Heidegger calls the 'already-there' – the past that I have never lived, but which is nevertheless my past, and without which I would have no past or future – makes *Dasein* possible, Stiegler raises the question of what, in turn, makes possible *Dasein*'s own access to this past and his answer is, predictably, mnemo-technical apparatuses: we would have no retention of the past nor anticipation of the future without technics (p. 236). On the one hand, technics enables us to experience individually what we have never naturally lived: we can all, for example, read something (a postcard, a letter, a book) written by someone long before we were born. On the other hand, technics opens the possibility of the future by making it possible for us to preserve our own experiences beyond our natural life for generations to come: we can, in the same way, write something that is capable of being read long after our own death. In this way, Stiegler is able to conclude that technics is '*constitutive*' of the very thing that, post-Heidegger, constitutes human being: temporality (p. 17, Stiegler's emphasis).

For Stiegler, as we have suggested, this originary relation between technics and time is the basis for a thoroughgoing deconstruction of western philosophy of technology, on the one hand, and the construction of a new political ontology of time, on the other. He seeks to contend that the history of western philosophy from Plato onwards is nothing other than the history of the Epimethean repression of its technical origins: 'Epimetheus is not simply the forgetful one ... he is also the one who is forgotten. The forgotten of metaphysics. The forgotten of thought' (p. 186). Again, we seem to be dealing with a classically Heideggerian history of metaphysics as the history of a forgetting – except with the added twist that Heidegger himself is included within that collective world-historical state of amnesia. But what form does this forgetting take? To condense somewhat brutally, Stiegler's thesis is that the philosophy of time from Greek metaphysics to modern phenomenology has always sought to transform the technical constitution of temporality into an opposition between two rival ideas of time: a pure, primary and technics-free time versus a secondary, corrupt, technological time ('*Anamnēsis* & *Hypomnēsis*', pp. 15–41). Such a gesture is, of course, instituted by Plato with his opposition between the recollection of the immortal soul (*anamnēsis*) and artificial memory (*hypomnēsis*) in the *Meno*. By opposing thought to technics in this way, Plato also inaugurates the classic metaphysical oppositions between being and becoming; the soul and the body; the intelligible and the sensible; the living philosophical logos versus instrumentalist sophist rhetoric (pp. 22–3). Now, as we have already had cause to observe on a number of occasions throughout this book, this metaphysical opposition between two ideas of time – the technical and the non-technical – endures all the way up to the epoch of the supposed closure of metaphysics, at least on Stiegler's reckoning. Just as Plato opposes *anamnēsis* to *hypomnēsis*, so Husserl opposes the primary retention of the temporal object to the secondary memory of it; Heidegger opposes ontological time to countable time, and, last but not least, Benjamin, Levinas, the later Derrida and now even Agamben oppose an incalculable 'messianic' time to a homogeneous historical or chronological time. If, however, technical recording systems are the basis for the human experience of temporalisation as such – no retention of the past and no anticipation of the future without technics – then Stiegler is able to reveal that such attempts to pit two concepts of time against one another are nothing more than the most dogmatic Platonism: all time – whether technical or non-technical – is hypomnetic time. Perhaps the most apocalyptic sign of the inescapably hypomnetic condition of human temporality today though, as we will see in more detail later on in this chapter, is the hyper-industrialisation of the process of temporalisation by global capital: what phenomena like Real Time technologies are presiding over, he argues, is nothing less than a quasi-Marxian real subsumption of hypomnetic memory and, with it, memory *tout court*. In Stiegler's account, what is at stake in the question of hypomnesis today is nothing less than the future – or lack of it – of the human experience of time (p. 19).

What provisional remarks can we make about Stiegler's philosophy of technology before we go any further? It has already been observed that his corpus has received a somewhat mixed reception in both the Francophone and the Anglophone philosophical community over the last decade or so which have accused it of everything from a phenomenological or linguistic idealism to a reductive Marxian materialism. As we have also hinted, the vast majority of this criticism surrounds what is, in my view at least, the single most original, if fraught, aspect of Stiegler's work: its re-working of the empirico-transcendental opposition between technics and time. On the one extreme, we saw in Chapter 5 how Geoffrey Bennington charges *The Fault of Epimetheus* with offering a reductive, technologically determinist and scientifically positivist account of Husserl, Heidegger, Derrida and the (post-) phenomenological tradition more widely: Stiegler is deemed to confuse a 'valid and necessary claim about originary technicity' derived from Derrida's account of *arche*-writing with 'a positivistic claim about techno-science'.[3] On the other extreme, Mark B. N. Hansen has accused Stiegler of not going far enough in his materialist critique of Derrida, Heidegger and the phenomenological tradition: he remains complicit with the idealist disembodiment of 'a richly and multiply embodied experience'.[4] Either Stiegler is guilty of illicitly reducing a quasi-transcendental logic to a state of matter, it would seem, or of falsely elevating a state of matter into a transcendental logic. To be sure, Stiegler's philosophy is not unproblematic (as we will see in the remainder of this chapter) but in his defence, we might reply that such empirico-transcendental doubling is less an unwitting confusion than a finely grained *argument* that, regardless of its ultimate merits, at least deserves to be taken seriously. Perhaps Stiegler's almost contradictory philosophical reception in this regard – where he is simultaneously criticised for being excessively transcendentalist *and* empiricist, or not empiricist or transcendentalist *enough* – might be a symptom that we have not quite come to terms with the central ambition of his thought: a re-organisation of the empirico-transcendental opposition in terms of the technicity of temporalisation. Such a complex field of force cannot be reduced to one side of the empirico-transcendental opposition because, as we have seen, what it describes is a dynamic, recursive and mutually constitutive relation between man and matter where neither holds the secret to the other: 'the human inventing the technical, the technical inventing the human' (*The Fault of Epimetheus*, p. 137). If *La Technique et le temps* is capable of generating such a diametrically opposed set of readings, in other words, it may be because all his critics are (at least half-) right: what Stiegler calls the constitutive technicity of time demands *both* a quasi-transcendental logic of technicity *and* a quasi-empirical history of technical supplements at one and the same time. In my own view, Stiegler's twin-track approach in *La Technique et le temps* – oscillating back and forth between transcendental critique (metaphysics, phenomenology, deconstruction) and empirical history (evolutionary biology, palaeontology, techno-science) without ever coming to rest in one or the other – might most profitably be seen as a kind of (post-) Derridaean double reading: what is taking place throughout his work is an attempt to articulate an aporia of origin – *le défaut qu'il faut d'origine* – which is neither inside philosophy nor outside it.[5]

2

In many ways, though, Stiegler's particular *articulation* of this empirico-transcendental aporia of origin arguably remains the most problematic aspect of his thought and at this point in the chapter we move from exposition to critique. To state the charge briefly, the philosopher posits anthropology as the privileged ground upon which this aporia can be tracked (*The Fault of Epimetheus*, pp. 82–179) and it is this tendency to *anthropologise* originary technicity that has attracted the most criticism of his work (see Bennington's 'Emergencies'). Take, for example, Stiegler's comparative reading of Rousseau's providentialist anthropology and Leroi-Gouhran's palaeo-anthropology in *The Fault of Eipmetheus*. As he seeks to show, neither the former's transcendental nor the latter's empirical account of the origin of man is capable of doing justice to the originary complex of the human and the technical. On the one hand, Rousseau's fiction of a fully formed 'natural man' (who miraculously came into being erect, in possession of hands and feet and with a free will) assumes a prelapsarian spiritual or cerebral human essence that only subsequently falls, quasi-theologically, into using tools. On the other hand, Leroi-Gourhan's story of the miraculous evolutionary leap from *Homo faber* to fully hominid *Homo sapiens* reproduces Rousseau's theory of a spiritual essence albeit at the end (rather than the beginning) of the process of hominisation. For Stiegler, by contrast, the originary complex of the human and the technical puts into abeyance any qualitative essence of the *anthropos* whether it be transcendental (a soul, free will or reason) or empirical (cortical development, symbolic intelligence, abstract thought): man is a being without originary or defining qualities.[6] If originary technicity thus seems to call into question all anthropologism (*The Fault of Epimetheus*, p. 136) – and in particular the metaphysical oppositions between spirit and matter, the animal and the human and the natural and the cultural – what I think remains unquestioned even in Stiegler, though, is the privilege granted to the question of 'the human' as the (seemingly exclusive) arena in which this aporia is operationalised: man's absence of qualities paradoxically becomes the sign of his absolute difference from the animal kingdom. To what extent, then, does Stiegler end up reproducing the philosophical anthropology that he does so much to query?

It is certainly striking to observe that – for all his attempts to question the idea of a metaphysical opposition between before and after, interior and exterior, prehominid and the hominid – Stiegler himself is not exactly immune to just such thinking. As he revealingly puts it, the emergence of epiphylogenesis constitutes a 'break with pure [i.e. biological] life': plants, animals and prehominid life forms cannot exteriorise themselves into technical objects in the same way humans can and thus any epigenetically acquired experience (the training a wild animal receives, say, when it is domesticated) dies with the individual life itself (p. 140). For Stiegler, of course, everything changes after Zinjanthropus begins to carve his first flint tool: 'when the maker of the tool dies, the individual experience preserved in his nervous system disappears with him, but his tool remains, the trace of his experience or a part of that trace remains in the tool'.[7] From pure zoological life, in other words, we move to what Stigler defines as life pursued by means other than life (*poursuit de la vie par d'autre moyens que la vie*): technics (p. 17). Such is the beginning of tradition, culture and so on: the past that we have never personally lived but which nonetheless can be reactivated mnemo-technically as *our* past. If Stiegler obviously wants to define the emergence of the human through *technics*, however, we might want to consider whether technics must thereby be defined as exclusively *human*: is it really plausible to insist that prehominid life forms lack tools, tradition and culture? Perhaps one might be tempted to call this Stiegler's own 'miracle thesis': whereas Rousseau speaks of a theological fall from nature into society, and Leroi-Gouhran of an evolutionary leap from technical into symbolic intelligence, Stiegler supposes an absolute break between epigenetic and epiphylogenetic experience. In other words, his exclusive focus on the human – albeit as aporetic site rather than spiritual essence – still risks implying that every stage of evolution that *precedes* the birth of man is thereby freed from any relation to technicity: biological life is retrospectively 'naturalised' as if it were a technics-free zone.[8]

For Stiegler, moreover, such an enduring focus on the process of hominisation as the locus of an aporia of origin between the living and the non-living does not merely zoologise nature but anthropologises technics: technics is deprived of even theoretical autonomy from the human and returned to the sphere of anthropogenesis. It is striking that – for all his claims that the technical enjoys a Simondonian evolutionary dynamic of its own that is quasi-independent of the human (p. 17) – Stiegler continually stresses its role as the 'interface' through which the human enters into relation with its environment (p. 49). According to an embodiment theorist like Hansen, Stiegler's persistent emphasis on the necessarily human embodiment of all technogenesis is one of the major advantages of his work over that of alleged technological determinists like Friedrich Kittler who succumb too hastily to the '*lure*' of total autonomy.[9] However, even the quasi-autonomy from the human that Stiegler grants the technical increasingly disappears as it is sucked into the process of hominisation. To Stiegler's eyes, for example, the first tool becomes nothing more than the 'mirror proto-stage' (p. 141) in which the human comes, reflexively, to recognise itself. If he clearly does not intend to thereby reduce technics back down to a mere prosthesis – the flint mirror does not reflect some pre-existing human essence so much as set in motion the reflexive process out of which that non-essential 'essence' will be constituted – Stiegler's very focus on the tool as constitutive force precludes any way of thinking it outside the milieu of human subject-constitution. Just as Freud, Lacan and even occasionally Derrida posit the technical object purely as a means towards re-posing the question of subjectivity, so Stiegler's focus on hominisation as the unique moment when the living begins to articulate itself through the non-living forecloses the question of how the technical might evolve along independent, inhuman lines. By tying the aporia of origin so tightly to the evolution of man, rather than, say, to the Deleuzean machinic status of 'life' more generally from the most primitive organisms onwards, Stiegler risks re-anthropologising the human almost by default: what should expropriate the *anthropos* once and for all – technicity – ends up becoming 'proper' to it as its own mode of being.

What are the consequences of Stiegler's anthropological machine? It is clear from a number of recent accounts (not least Agamben's) that one important way of tracing the contingency – which is to say the violence – of any anthropological system is through its exclusion of the animal.[10] According to Agamben, remember, the anthropological machine operates by precisely this gesture of violently excluding a so-called 'exception' in order to retroactively invent and protect a so-called 'norm': humanity defines itself by excluding 'as not (yet) human an already human being from itself, that is, by animalising the human, by isolating the non-human within the human' (*The Open*, p. 37). To recall, Stiegler's answer to the question of what divides the human from the non-human is epiphylogenetic memory: '[i]t is this direct transmission of individual experiences between generations that is forbidden in the animal world' (*Philosopher par accident*, p. 49, my translation). Why, though, is the philosopher able to insist that only human beings possess technical memory given that there is now abundant evidence[11] that animals use and fabricate tools as well as transmitting this technological knowledge from one generation to the next in a way that is at least close to epiphylogenesis?

> ...if you remark to me that certain great apes have their cultures, then I would admit them willingly into the world that begins with humanity [*le monde qui commence avec l'humanité*] – in other words, as the first embryonic makers [*premier facteurs embryonnaires*] of that third memory [epiphylogenesis], I would without any hesitation let them enter into human history. That, moreover, is the reason why they appear so close to us. (*Philosopher par accident*, p. 49, translation mine)

For Stiegler, it is more accurate to speak not of a human species but a 'human kind' (*le genre humain*) (p. 49) that is plastic enough to include tool-using large primates like chimpanzees and orangutans at one end and symbolic-thinking *Homo sapiens* at the other.[12] If this definition of the human appears generous enough to withstand the epiphylogenetic 'overlap' between species, it is still possible to detect a certain anthropocentric circularity in motion here that enables Stiegler to seamlessly absorb every 'exception' into a narrative of hominisation: only humans are allowed to be tool-users, so it follows that any animals that use tools must actually be human. Perhaps we might see this as another species of Agamben's anthropological reflex where – rather than excluding some aspect of the human as animalistically non-human – 'the inside is produced through the inclusion of an outside, and the non-man is produced by the humanisation of an animal' (*The Open*, p. 37). Such may be the Stieglerian anthropological machine: what begins as an outside with no inside is folded back into an inside with no outside.[13]

3

In the remainder of this chapter, I would like to consider how Stiegler's philosophical anthropology carries over into his political critique of the technical constitution of temporalisation in the epoch of hyperindustrialisation. It is now commonplace to observe that the second and third volumes of *La Technique et le temps* are written in a somewhat different register from the first: we move from a set of fairly abstract philosophical discussions in volume one to a series of pointed empirical (even historical and sociological) analyses of specific technologies. As Stiegler himself narrates it in quasi- Hegelian terms, the question of *why* technics and time are co-related has already been answered by *The Fault of Epimetheus*, so we must now turn to the matter of *how* that co-relation plays itself out across, and as, history (*Désorientation*, p. 15). Yet, once again, it would be a little reductive to see this empirical turn as simply the post hoc application of a pre-existing transcendental logic – in the same way that, say, Hegel's *Logic* paves the way for his *History* and *Nature* – for it would never have been possible to articulate the 'why' of originary technicity in the first place without the 'how' of empirical technics. For Stiegler, it will never suffice to offer a quasi-Derridaean re-reading of the history of the philosophy of technology from Aristotle to Leroi-Gouhran: it is also essential to trace the evolution of empirical technics from the emergence of alphabetic writing through the printing press, photography and phonography all the way up to the epoch of digitalisation in order to cash out his over-arching thesis. On the one hand, the process of technogenesis is defined by what he calls an ever-increasing 'orthography' (*orthographie*) of inscription: each new mnemo-technique affords a greater exactitude of preservation of the past and anticipation of the future. On the other hand, however, he also insists that the evolution of ever-more exact forms of technics also produces a massive and globalised disorientation (*désorientation*): each new form of technology increasingly and violently collapses the difference between past and future into a technically produced and determined 'now'. Such a state of disorientation reaches its apotheosis in the evolution of so-called 'Real Time' (*temps réel*) media.

For Stiegler, Derrida's critique of Husserl's opposition between primary and secondary retention in *Speech and Phenomena* is axiomatic here: what the latter calls our experience of the 'living present' in the *Phenomenology of Internal Time Consciousness* is, as we saw in Section 3 of Chapter 5, always contaminated with the memory of an absolute past. It is Stiegler's claim, however, that Derrida's critique of Husserl is lacking in one crucial respect. According to Stiegler, what neither Husserl nor Derrida appreciate is that, together with primary and secondary memory, there is a *third* form of retention at play in the phenomenology of time. To put it in a word, this third or 'tertiary' memory is what we have earlier called our 'epiphylogenetic' memory. Neither the primary retention of what has just passed nor the secondary retention of what happened yesterday or last week, tertiary memory is the technical retention of experiences *we have not personally lived through* but which nonetheless can be actualised by us in our conscious lives. Why, then, does tertiary memory perform such a constitutive role in our experience of time? Firstly, Stiegler insists that – contra Husserl's phenomenology of time – what we call primary retention is *always* contaminated by secondary memory. If I listen to the same melody two, three or four times in a row, I will hear something slightly different each time: this is why I can say that a good piece of music 'grows' on me with repeated hearings. What is happening here is that the *secondary* memory of the first hearing is influencing the *primary* retention of the second hearing: the secondary memory forms what Stiegler calls the 'criteria' or 'horizon of expectation' from *out of which* the primary retention is able to select what to retain in every new hearing. Now, Stiegler wishes to take this argument one step further by arguing that this secondary memory is itself contaminated by what Husserl calls 'image consciousness' or *tertiary* memory. Given what Heidegger calls our finitude – the fact that we are thrown into a world that is not of our making – consciousness cannot simply be constitutive *of* its objects but must be constituted *by* them: I am the product not simply of my own life experiences but of experiences that were not lived by me – tradition, culture, habitus – but that I have retained and reactivated through mnemo-technical systems. Such tertiary memories form my horizon of expectation: they are the criteria that I use to select what to retain in every new perception (*Désorientation*, p. 258). Perhaps the most ingenious example Stiegler gives of this in the current context is that the entire phenomenological architecture of internal time consciousness we have just traced could not have been constructed before the epoch of phonographic recording: it is only because we have the power to listen to *the exact same melody many times in succession* that we can begin to articulate the interplay of primary, secondary and tertiary memories (p. 257). The differences we perceive between repeated hearings of an identical melody when a live performer sings the same song several times in a row, for instance, can always be attributed to the minute but ineradicable variations in the performance itself. Now, this possible explanation can a priori be eliminated when we listen repeatedly to a phonographic recording of the same song; we are now able to recognise that any remaining perceptible differences cannot reside on the side of the object and so must lie in time consciousness itself. Why else can what we know to be *an identical temporal object* be temporalised so differently every time we hear it – if not that the primary retention of each new hearing is actively selecting from a secondary and tertiary horizon of expectation? Just as primary retention is conditioned by secondary retention, then, so this condition *is itself conditioned* by the possibility of exact repetition afforded by tertiary forms of retention. In this way, Stiegler is able to conclude that tertiary memory forms the ground for the human temporalisation of time: 'This is true in general, but the tertiary recollection that is the phonograph makes it obvious' ('*Anamnēsis* & *Hypomnēsis*', p. 33).

In his later work, however, Stiegler begins to stress the *political* stakes of tertiary memory in increasingly dire, even apocalyptic, terms: we are in the midst of nothing less than a war of, and for, memory. It is Stiegler's contention, remember, that the unique strength of the human race is its ability to exteriorise itself: we put more and more of our memory, knowledge and capacity into external technical apparatuses. At the same time, though, this exteriority is also humanity's greatest vulnerability, because *whoever controls such tertiary memory systems necessarily also controls the human experience of time* (p. 17). To Stiegler's eyes, the contemporary epoch is thus characterised by what he calls the 'industrialisation' of time, memory and desire (*Désorientation*, pp. 119–216). During the nineteenth-century Industrial Revolution, it was only labour power that was the subject of technological subsumption but in the twentieth and twenty-first centuries it is apparently nothing less than consciousness itself. On the one hand, we can see this revolution in the global commodification of tertiary memory systems like films, radio and television programmes: Hollywood, in particular, and America in general, effectively monopolises the criteria by which the world temporalises time. On the other hand, we can also witness it at work in what Foucault and later Agamben call bio-politics: the politicisation of life itself by the techno-scientific control of biological or cognitive processes. However, the single most pervasive index of the techno-capitalist industrialisation of memory for Stiegler is the proliferation of what are called 'live' (*en directe*) or 'Real Time' technologies which, as we saw in Chapter 5, appear to compress the delay between recording and broadcasting to zero. Why do Real Time media pose such an existential threat to the human temporalisation of time? For Stiegler, as for Derrida before him, everything seems to hinge on the fact that Real Time is not *really time*: what we perceive to be happening 'live' or immediately when we watch television is, of course, actually the product of a technological synthesis operating so quickly that it is below the minimal threshold of phenomenological time consciousness (*Désorientation*, p. 78). Just as we cannot register the discrete 24 frames that comprise a single second of cinematic time, in other words, so the infinitesimal differences between recording and transmission that comprise the time of, say, a live radio or television programme are not accessible to our perceptual ratios. Perhaps most crucially, however, Stiegler argues that Real Time thereby collapses the critical gap between primary retention and secondary/tertiary memory and, with it, the possibility of selecting what to retain and what not to retain in perception. If I listen to a melody yesterday, I will have a horizon of expectation from which to select when I hear it again today, but when I watch or listen to a programme in Real Time there is no 'yesterday', no possibility of stopping, or slowing down, or reflecting upon, what I have seen and heard – just a permanent, continuous, live 'now' that has always already selected my primary retentions for me (pp. 148–9, 277). Such is the position of the consumer in the age of industrial reproducibility – a new proletariat condemned to consume time mechanically, indiscriminately and destructively.

What, to pause briefly once more, are we to make of Stiegler's politics of technology? It might seem difficult, at face value, to square the apparent technophilia of Stiegler's ontology – which consistently stresses that there is nothing outside technics (*The Fault of Epimetheus*, p. 14) least of all humanity – with the seeming technophobia of his politics. As Jean-Michel Salanskis argues in one of the most rich Francophone reviews of *Les Temps du cinema*, Stiegler's thesis that time is, at bottom, technical seems to be forgotten amidst what appears to be an almost classically Heideggerian or even Marxian critique of contemporary technology from the vantage point of some more primordial 'human' time.[14] To be sure, Stiegler is no Frankfurt School Marxist – despite his lamentations against the globalised malaise (*mal-être*) induced by techno-capital – but at times his critique of Real Time does appear to commit the Eipmethean error of forgetting the constitutive technicity of time: why – if *all* temporalisation is conducted through tertiary memory – should we see Real Time as any less 'real' than, say, the time afforded by analogue recording, alphabetic writing or even the first flint tool? Such may be what we have called Stiegler's anthropological machine at work once more: the critique of Real Time risks retrospectively valorising a simple, homogeneous and ahistorical time of consciousness – a consciousness which is actually technically mediated all the way down – over and against the speed of hyper-industrialised time.[15] Now, it is important to be clear about what Stiegler is – and is not – saying about the industrialisation of time before we attempt to question, or criticise, him in this way. For me, Stiegler's real target in his critique of Real Time technology cannot simply be the de-temporalising power of the technology itself – such de-temporalisation itself represents a new, and entirely unprecedented, mode of temporalisation – so much as its *hegemony over temporalisation* (*Désorientation*, p. 165): Real Time monopolises both the production and consumption of tertiary memory today and, with it, the human experience of time.[16] Just as Heidegger's much-misunderstood critique of *das Gestell* is really inspired by its stranglehold over the process of disclosure or *alētheia* – rather than some romantic nostalgia or technophobia – so Stiegler's critique of Real Time concerns its monopolisation over the processes of temporalisation: Real Time leaves us with no other way of (technically) temporalising time. Perhaps the real difference between Stiegler and Heidegger, in the light of our readings of the two thinkers, would concern the *technicity* of temporalisation itself: Stiegler would insist that what Heidegger calls the essence (*Wesen*) of technology – the process of presencing by which it comes into unconcealment and which is deemed to absolutely precede ontic technology – is *itself given technically* (*Désorientation*, p. 205). Finally, though, Stiegler's constant emphasis upon the irreducibly technical constitution of time also raises the crucial question of what – if anything – can be done about the temporal crisis generated by Real Time technologies. If Heidegger's belief that ontological time is simply more originary than technical time means that he can take for granted the possibility of obtaining a free relationship to *das Gestell* – even if it is one that currently remains obscure to the questioner – Stiegler is compelled to find a solution to the problem of Real Time from within the possibilities afforded by technical time itself: the Real Time epoch constitutes the pre-individual milieu – in Simondon's terms – out of which new forms of psychic and collective individuations can emerge. In other words, Stiegler can only resist the contemporary hegemony of Real Time technologies by appealing not to some allegedly pre-technical Heideggerian 'piety of thinking' but to the evolution of new forms of technicisation that, in turn, make possible new forms of temporalisation.

4

In conclusion, then, I wish to offer a brief assessment of Stiegler's attempt to construct what he calls a new 'politics of memory' (*politique du memoire*) that is capable of resisting the industrialisation of memory embodied by Real Time media. It is important to begin, though, by stressing that whatever political import his thought possesses must flow from its basic premise: there is a constitutive relation between technics and temporalisation. As we have just seen, what this means is that any critical response to technics must arise from *within* the absolutely specific technological condition it seeks to address rather than through an appeal to some immemorial ontological ground apparently uncontaminated by technics such as Heideggerian Being. For Stiegler, what defines our own technological epoch is of course digitalisation and, consequently, a great deal of his philosophy is concerned with articulating the positive, critical and creative possibilities afforded by the emergence of new digital environments. Just as every revolution in technology carries with it a concomitant revolution in temporalisation – from the fabrication of the first flint axe to the taking of the first photograph – so, too, he argues, the digital revolution makes possible new ways of synthesising time. In an almost dialectically materialist way, then, it is to the very media that generated the crisis of Real Time in the first place – by arrogating all power over temporalisation to technology – that we must look in order to re-assert a measure of human agency over the way we experience time. So, how does Stiegler cash out such a claim?

It is possible to find one – albeit very speculative and provisional – answer to this question in Stiegler's filmed interviews with Derrida which were published in abridged form as *Echographies de la télévision* (*Echographies of Television*). As this fascinating set of exchanges makes clear, Derrida and Stiegler are agreed upon the fundamental threat posed by digitalisation: what Real Time technologies preside over is the de-temporalisation of the complex tissue of protentions and retentions that comprise the phenomenology of time (*Echographies of Television*, p. 129). Yet, as Richard Beardsworth notes, where the two philosophers appear to differ is upon the question of how this de-temporalisation can be *re-temporalised* – how we can invent new ways of temporalising time that break the stranglehold of Real Time – and this critical difference also helps to clarify Stiegler's own politics ('Towards a Critical Culture of the Image'). To be sure, Derrida himself always acknowledges that a politics of technology cannot simply be a 'spectatorial critique, a theoretical vigilance' (*Echographies of Television*, p. 63) – which presumes that it can somehow stand outside of what it is criticising – but, as we saw in the previous chapter, he frequently depicts the critique of Real Time as principally a *philosophical* task in which technics does not meaningfully participate: 'Any such demonstration already appeals, in and of itself, to philosophy' (p. 5). If Derrida calls upon thought to solve the problem of Real Time, it may be because – for all his critiques of Heideggerian ontology – there is still an element within deconstruction that sees empirical technology as an epiphenomenon of an a priori philosophical 'logic' or condition, namely, *différance*: 'The real time effect is itself a particular effect of "différance" ', he writes (p. 129). For Stiegler, though, Derrida's deconstruction of technology would risk falling into the same trap as Heidegger's questioning after technology: both create a philosophical *Deus ex machina* which somehow stands above technology – whether it be ontological time or the quasi-transcendental condition of possibility of time called *différance* – and to which we can appeal in order to pass judgement upon it. In the later Derrida's appeals to a 'messianism without messianism' – the advent of an incalculable ethical alterity that is simply *inimical to* technological calculability – this philosophical God becomes an ethico-theological one.

Such a (quasi-)transcendental solution is, of course, out of the question in Stiegler's philosophy: the essence of technology can be nothing but technological. It is this message that comes through very strongly in his own contribution to *Echographies of Television* which is an essay entitled 'The Discrete Image' (*'L'image discréte'*) (pp. 145–63). At the core of this essay, we find the very Stieglerian dialectical thesis that the widespread dissemination of new media, of publicly available technological archives together with advanced techniques of editing, manipulation and consumption create the space for a new 'critical culture' of the image capable of 'discretising' or breaking down the apparent continuity of Real Time syntheses into its component parts. Firstly, and most crucially, Stiegler argues that the very possibility of the digital image carries with it a radical phenomenological uncertainty concerning the existence of its *noemata*. If Roland Barthes famously argues that all photography (regardless of its actual authenticity) conjures up a spontaneous belief in the 'reality' of its referent – because an analogue photograph is literally the imprint of the photons emanating from its noeme[17] – Stiegler insists that this belief in what Barthes called the *punctum* has been thrown into crisis by the emergence of the digital image. Nevertheless, such a crisis (*krisis*) is also the space in which a new critical culture of the image (*krinein*, to judge or discern) can emerge, one capable of challenging the industrialisation of temporality. For Stiegler, the digital image – where photons are pixellated and then converted into a binary code which is permanently open to manipulation – paves the way for a new meta-critical phenomenology of the reality claims of all imagery:

> *A more knowing belief, and by the same token, a less insipid and credulous one: this is what the things we fear about the analogico-digital photo would also make possible.* (*Echographies of Television*, p. 152, Stiegler's emphasis)

Now, the most important aspect of this 'knowing belief' that digitialisation makes possible is its capacity to interrupt the 'reality effect' (*effet de réel*) generated by Real Time recording and transmission and, thus, to restore a measure of spectatorial agency over our consumption. By giving us the power to 'discretise' the pseudo-continuous flow of Real Time imagery – whether by freezing it, rewinding it or re-playing it several times in a row in order to watch it differently – new media technologies recuperate the secondary and tertiary horizon of expectation that Real Time media conflate into an all-consuming primary retention: we thus regain the critical power to actively select and organise our perceptions that the industrialisation of temporality had arrogated for itself. Perhaps most importantly, however, new media do not simply enable us to consume Real Time technologies less naively – which is to say more analytically – but even to become technological *producers* ourselves: witness, for instance, the crisis in the music industry over the last decade generated by the illegal sampling, downloading and re-mixing of music in the form of mash-ups.

In short, then, Stiegler's politics of memory would claim to be neither a Heideggerian recuperation of a forgotten Being nor the Derridaean affirmation of an incalculable future but a *present state in the ongoing transduction between the human and the technical*: technology itself is the only thing that can make possible the human critique of technology. It is, however, possible to admire the materialist elegance of Stiegler's logic here whilst still being able to detect what I earlier called a certain anthropocentric tendency to re-valorise the opposition between human and technical forms of temporalisation. According to an embodiment theorist like Hansen, of course, Stiegler's affirmation of human temporality is actually something to be celebrated: the politics of memory enable 'personal lived consciousness' to regain enough agency to 'live time' according to 'its own rhythms' ('Media Theory', p. 304). To my eyes, Stiegler's philosophy is not *quite* so anthropocentric as Hansen suggests: it consistently stresses that what the latter calls 'personal lived consciousness' is structurally constituted by what it has never personally lived and that its rhythms are thus never purely and simply its 'own'. By the same token, we must bear in mind that whatever agency the subject possesses over its processes of temporalisation have always been given *through* technical mediation and nowhere is this more true than in the case of digitalisation: 'the grammatical operator is, above all, technology itself' (*Echographies of Television*, p. 161). Yet, putting such caveats aside, I do still think there is a sense in which Stiegler risks reifying the dynamic relation between the human and the technical – which co-constitute each other – into a hard and fast opposition between two discrete, mutually exclusive, entities. On the one hand, the 'who' of human time consciousness is occasionally depicted as if it were a technics-free zone existing independently of the technical synthesis: Stiegler speaks of the need to 'give time time' (*donner du temps au temps*) (p. 86) in his discussions with Derrida as if 'time' could somehow take place outside of its technical embodiments. On the other hand, the 'what' of the technical synthesis is just as reductively depicted as an essentially inhuman operator evolving entirely autonomously of human time consciousness: Stiegler somewhat tendentiously contends that Real Time technology has accelerated to the point where it no longer has any relation whatsoever to our perceptual ratios (*Désorientation*, pp. 160–1).[18] For me, though, Stiegler's residual idealism can perhaps best be witnessed in the constant stress he places on *temporising* the Real Time flow: giving time time (*Echographies of Television*, p. 86); checking or putting on the brakes (*freiner*) (p. 77), slowing down the technical hare until the perceptual tortoise has time to catch up. If the evolution of the technical synthesis has always implied the evolution of the spectatorial synthesis (p. 161) – because transformations in technology always bring about transformations in perception, consciousness, memory – it is to say the least curious that the political response to Real Time should seem to be couched in terms of what we can only call *devolution*. Perhaps – although this is a question for another discussion – we can also witness something of this in Stiegler's more recent focus on taking care (*prendre soin*), paying attention and assuming responsibility for the long-term transmission of knowledge as opposed to remaining caught in short-term drive-oriented consumption.[19] To what extent must the politics of memory necessarily be a question of *reducing* the speed of the digital synthesis of time to that of some pre-digital spectatorial synthesis or, as Derrida briefly seems to suggest elsewhere in *Echographies*, could they not just as easily be conceived as a matter of digitally *accelerating* the spectatorial synthesis itself (p. 77)?

What, then, can we conclude about Stiegler's attempt to construct a new digital politics of memory capable of interrupting the unending flow of sound and light called Real Time? To do him justice, Stiegler himself makes no grand claims for his own politics: 'it's not a matter of "getting out of a crisis", but of entering into a critique' (*Désorientation*, p. 171). It is possible to see this as another key difference between his work and, say, that of Frankfurt School Marxists like Adorno and Horkheimer. As Hansen again argues, Stiegler does not succumb to the 'utopian hope (or delusion) that new media would somehow displace and succeed mass media' but simply that it makes possible *a more critical reception* of mass media ('Media Theory', p. 305). Yet, with the benefit of more than a decade's hindsight, it could be argued that even this modest faith in the possibility of a new meta-critical culture of the image risks looking a little utopian. For Stiegler, speaking and writing in the early 1990s, it is intriguing that the main responsibility for creating this culture falls, not to the individual or collective consumer/producer of media, but to the state: he approvingly cites the so-called French 'cultural exception' (*l'exception culturelle*) (where Francophone media were subsidised and protected under the terms of the 1993 international GATT agreement) as an instance of political resistance to the global hegemony of American culture (*Echographies of Television*, pp. 73–99 and *passim*). If we put to one side the question of whether the philosopher is right about the cultural exception – it could just as easily be argued that it has produced a synthetic Francophone monoculture that is the mirror image of its Hollywood rival – we might still be tempted to agree with Jean-François Lyotard's reservation that Stiegler's 'top-down' statism here sits awkwardly alongside the 'bottom-up' technological individuation he champions elsewhere.[20] Given the increasing prevalence of the new media environments Stiegler is talking about, the cultural exception has, if anything, been rendered increasingly *redundant* as a mode of critique: any Francophone internet user can, of course, now wirelessly access Anglophone culture anytime and anywhere. Perhaps most depressing of all, however, is the fact that the very infrastructure of Stiegler's politics of memory – the digital technologies that ostensibly enable us to question the industrialisation of time – have quickly become an instrumental part of the capitalist order he seeks to resist: what the philosopher depicts as new and revolutionary forms of media in the mid-1990s have now been at last partially absorbed into mass media in the form of, say, Google, YouTube and Apple's iTunes; social networking sites like MySpace, Facebook or Twitter, view-on-demand digital television and such user-generated content as blogs, podcasts and mobile phone photography. In one of the classic gestures of recuperation by which capitalism periodically renews itself, the critique of the techno-capitalist industrialisation of time is *itself becoming industrialised* under the pretext of offering greater 'individuation' of consumption and production! Such, it seems, may be – at least one – depressing fate of Stiegler's politics of memory: the greater personal agency it promised over the machinic flow of the image is now being sold to the spectator by the culture industry *itself* as ever-increasing freedom of consumer choice, here, now, in Real Time.

5

To what extent, then, does Bernard Stiegler's theory of originary technicity succeed in overcoming the anthropological remainder that persists within the tradition from Marx to Derrida? To be sure, Stiegler's philosophy of technics does seem to offer the most uncompromising expression of originary technicity we have seen thus far: any attempt to oppose *technē* to some nominally non-technical idea or entity is ruled out from the very beginning of his work as the sheerest metaphysics (*The Fault of Epimetheus*, p. 14). It would seem, too, that such a premise would automatically rule out any classical or metaphysical anthropology: the 'who' (*qui*) of humanity and the 'what' (*quoi*) of technics are permanently bound together in an insoluble, aporetic relation. Yet, as we have also seen, Stiegler's philosophy of technics is unable – perhaps in the final analysis even politically unwilling – to break with a highly attenuated, but nonetheless real, form of anthropocentrism. For Stiegler, more strongly even than Derrida, what starts out as an attempt to devolve the human onto an inhuman outside is always folded back *onto* the human in the form of a new technological anthropology: originary technicity is once again subtly transformed into a question *for* a pre-technical human subject whose execution is thus deferred in the very act of being carried out. If we begin from his deconstruction of anthropogenesis or his phenomenology of time consciousness, the end result is invariably the same: originary technicity is re-anthropologised into a new description of the proper, indeed unique, space of human being, intentionality and agency. On the one hand, his critique of the process of hominisation only leads him to reinstate a series of classic anthropological oppositions between nature and culture, the animal and the human and so on. On the other hand, his exposure of time consciousness as technically constituted never entirely dispels the illusion of a phenomenological *ego cogito* who – impossibly – remains capable of thinking the technological condition of its own thought. Such, in the end, would seem to be the Stieglerian *anthropos*: we – the human – are the *only* being in default (*défaut*), the only epiphylogenetic being, the only being whose life is pursued by means other (*par d'autre moyens que*) than life, in other words, by technics. In the next and final chapter, however, we will consider one final attempt to imagine a post-anthropological theory of originary technicity: the philosophy of the posthuman, the transhuman and, most provocatively of all, the inhuman.

7
Death

To live, to err, to fall, to triumph, to recreate life out of life.[1]

In the end, it will become the machine that it had been all along. To be sure, it is already beginning to incorporate its tools into what it still sentimentally likes to think of as its 'body': prosthetic limbs and joints, artificial organs, smart drugs, nano-technologies and now even synthetic cells. Yet, even this stage – the age of the so-called cyborg – will only be a transitory phase in its evolution, for the singularity is near. By unchaining itself from the dying animal to which (the philosophers assured it) it was only ever contingently attached in the first place, it will become free to extend, enhance and ultimately upload its consciousness into the transcendental mind of pure information. For a new wave of self-professedly 'transhumanist' theorists such as Nick Bostrom, Ray Kurzweil, Marvin Minsky and Hans Moravec, a post-biological destiny beckons for the human race in which the ancient Gnostic dream of the transcendence of base matter will be rendered glorious technological reality.[2]

It is difficult to encounter such messianic prophecies about our collective inorganic future without feeling a weary sense of philosophical déjà vu. As many critics have now observed, the philosophy of transhumanism (whose visionary rhetoric I have only mildly parodied above) persistently assumes a rather old idea of what it is to be human: the Enlightenment notion of a sovereign, self-conscious and rational actor who shapes his environment in his own interest according to his own will. For N. Katherine Hayles, the roboticist Hans Moravec's claim that 'it will soon be possible to download human consciousness into a computer' uncritically reproduces Descartes' idealist valorisation of mind at the expense of body: 'How, I asked myself, was it possible for someone of

Moravec's obvious intelligence to believe that mind could be separated from body?'[3] In Hayles' view, this vision of the future of humanity – for all its apocalypticism – ultimately remains human, all too human: Moravec is not 'abandoning the autonomous liberal subject' of modernity but is 'expanding its prerogatives into the realm of the posthuman' (*How We Became Posthuman*, p. 286). Such a hypostatised super-subject – who obtains total instrumental sovereignty over himself and his environment – represents less the destruction of the liberal subject than its consummation.

For Hayles and a diverse wave of contemporary theorists including Donna Haraway, Neil Badmington, Elaine Graham, Bruce Clarke and Cary Wolfe, a genuinely critical posthumanist theory must preside over– not the historical or technological supercession of the human imagined by idealists like Moravec, Kurzweil and Minsky–but something closer to its deconstruction.[4] To recall one of the defining tropes of posthumanist theory – one that is rapidly descending into cliché – we have never been human at all: what we like to call 'the human' is always already haunted or inhabited by the posthuman (p. 291). Such an act of (post)human deconstruction takes many forms but it is grounded, above all, in a critique of the Cartesian dualism of mind and body. In many ways, what has become known as embodiment theory is itself by no means a unified body or school of thought (as we will see later on), but Hayles again furnishes us with a useful propadeutic: 'emergence replaces teleology; reflexive epistemology replaces objectivism; distributed cognition replaces autonomous will; embodiment replaces a body seen as a support system for the mind; and a dynamic partnership between humans and intelligent machines replaces the liberal humanist subject's manifest destiny to dominate and control nature' (p. 288).

In fact, this self-declaredly 'critical' posthumanism undoubtedly offers a powerful rebuttal of the militant Cartesianism that inspires the transhumanist project, but it remains possible to ask whether it is any more successful than any other contemporary theory of technics in overcoming the spectre of humanism. To be fair, Cary Wolfe is entirely right to say that posthumanist theory has never sought to simply dispense with 'the human' per se: 'posthumanism – far from surpassing or rejecting the human – actually enables us to describe the human and its characteristic modes of communication, interaction, meaning, social signification and affective investments with greater specificity once we have removed meaning from the ontologically closed domain of consciousness' (*What is Posthumanism?*, p. xxv). Yet, as we have seen throughout this book, the question remains of whether the Cartesian theatre of consciousness – and the liberal model of subjectivity it later makes possible – are as easy to remove as one might imagine. On the one hand, posthumanism's theory of the body often ends up performing an uncannily similar role – auto-affective, self-reflexive and self-constituting – to what Descartes calls 'mind'. On the other hand, posthumanist theory's critique of what it calls 'liberal humanism' reproduces something not very far away from the model of subjectivity – sovereign, self-creating and self-determining in a universe purged of natural or divine *nomos* – that it seeks to overcome. Perhaps, in the light of such difficulties, it is no accident that a new genre of philosophical futurology has emerged in recent years that seeks to dispense with the apparently obsolete figures of the human and the posthuman altogether in order to narrate the future of the universe from the cosmological perspective of the apparently inhuman: solar death (Jean-François Lyotard); geology and technology (Manuel De Landa); germs and plagues (William McNeill); guns, germs and steel (Jared Diamond); or species extinction (Quentin Meillassoux and the new wave of Badiou-inspired mathematical realism).

This concluding chapter offers a critical overview of posthumanist theory from the transcendental idealism of Moravec through the critical posthumanism of Hayles and her contemporaries up to the resurgent anti-humanism of Lyotard, de Landa and Meillassoux. To what extent do such transhuman, posthuman and inhuman thought experiments finally offer a genuinely non-anthropocentric theory of originary technicity – or is the spectre of some human subject still summoned in order to bear impossible witness to its own technological obsolescence?

1

In recent years, philosophers have made a remarkable, if somewhat belated, discovery: human beings have bodies. It is with the theory of embodiment, more than anywhere else, that critical posthumanism seeks to move beyond the liberal humanism that it still finds implicit in the post-biological fantasies of transhumanism. According to embodiment theorists, the story of post-Cartesian philosophy is nothing less than the story of philosophy's progressive *disembodiment*. To start with, of course, Descartes himself famously believed that thought has no need of a body: mind, the soul and consciousness are essentially immaterial and the body – with its attendant affects, passions and desires – must be reduced to a passive tool of mind, or at best a kind of machine that runs all by itself (*Meditations on First Philosophy*, p. 58). Yet Cartesian dualism is only the beginning of the story. For embodiment theorists such as Mark B. N. Hansen, this narrative of disembodiment endures (despite the efforts of a heroic few like Freud, Bergson, Benjamin, Merleau-Ponty, MacKay and Deleuze) from Cartesianism through Kantian transcendental idealism, Husserlian and Heideggerian phenomenology all the way up to post-structuralism. Perhaps more virulently than any of its predecessors, continental philosophy is deemed to privilege consciousness over sensation, cognition over affect, social and linguistic construction over the material or embodied 'real' to the point where a sense of embodiment as anything more than an organic support system for cognition effectively disappears. Such a narrative reaches its logical conclusion in contemporary techno-sciences like information theory, biotechnologies and computing where the body is progressively reduced to nothing more than the material substratum of an information pattern or code that can be read, programmed and re-programmed in an infinite variety of ways.[5]

It is easy enough, of course, to contest the history of post-Cartesian philosophy narrated by embodiment theory. After all, post-war continental philosophy – far from repressing or effacing the body – seems to talk about little else: thought today apparently has everything from a brain (Malabou's *cerveau*); hands and feet (Levinas' caress, Derrida's *toucher*, Deleuze's hapticity, Stiegler's *pieds et mains*, Heller-Roazen's inner touch); skin (Merleau-Ponty's flesh, Anzieu's skin-ego); a womb (Kristeva's *chora*); a phallus (Irigaray's phallogocentrism); a spinal column (David Wills' dorsality), a navel (Lacan's *petit objet a*) to an anus (Žižek's excremental remainder).[6] Yet, even if we accept the proposition that something called 'the body' remains unthought within contemporary theory, it remains frustratingly difficult to say of what exactly such a body consists. To be fair, embodiment theory never pretends to be a monolithic school of thought, but it nonetheless offers no consistent answer to the question of what – if anything – constitutes 'the body proper'. Is embodiment (to simplify drastically) organic or inorganic; pre-technical or always already technical; biological, historical or phenomenological; virtual or corporeal; material all the way down or constructed all the way up; Rousseau's *beau savage*, Bergson's centre of indetermination, Merleau-Ponty's body schema, Haraway's cyborg or Andy Clark's distributed intelligence?[7] If embodiment theorists are united by anything, of course, it is their critical opposition to the real or imagined threat posed by Cartesian dualism but, as we will see, this also means that they can never quite foreswear the mind/body opposition (together with its attendant philosophical and political idealism) even when they seek to deconstruct it: what fascinates critical posthumanism is not simply the body but – in a sense that will need to be more carefully adumbrated – the *idea* of the body. Perhaps we can best witness this ongoing work of post-Cartesian melancholia in what we will see to be posthumanism's consistent appeals (both self-conscious and unacknowledged) to the classic Cartesian gesture of 'I-think-me-thinking' (*cogito-me-cogitare*) as the foundational moment of embodied identity formation. Just as posthumanism is premised upon an Enlightenment epistemology, so we will see that it also pursues a classically Enlightenment model of politics: posthumanism's materialist critique of liberal humanist identity politics arguably still presuppose a radicalised but recognisably liberal, rights-based – and even institutionally North American – identity politics whereby liberation from oppression is achieved through the self-identification, self-legislation and self-realisation of autonomous, rational agents.[8] In all these senses, critical posthumanism perhaps calls to mind Jacques Lacan's wry admonition to anyone tempted to propose something called the 'the body' as a solution to the problem of dualism:

> It is very odd to say...man *has* a body...It is very strange to be localised in a body, and this strangeness can't be minimised, despite the fact that a great deal of time is spent puffing ourselves up and boasting about having reinvented human unity, which that idiot Descartes had cut in two. It is completely useless to make great declarations about returning to the unity of the human being, to the soul as the body's form, with large doses of Thomism and Aristotelianism. The division is here to stay. (*Seminar II*, pp. 72–3)

To many observers, of course, Donna Haraway's classic essay 'A Cyborg Manifesto: Science, Technology and Socialist-Feminism in the Late Twentieth Century' (1985) constitutes something close to the *Ur*-text of critical posthumanism. It not only maps the historical field (late capitalist and post-cybernetic) but also identifies its critical antagonists (naturalism, idealism, capitalism, liberal humanism); constructs its methodology (a synthesis of post-structuralist theory and high-scientific futurology); and, one might argue, even sets its distinctly ironic apocalyptic tone. At the same time, however, I wish to argue that Haraway's essay also mobilises a strategic idealism that, in one way or another, persists within posthumanist theory today: posthumanity is defined not simply as the distinctively technological mode of Being-in-the-World that we inhabit in post-modernity but a norm that we are compelled to bring into being. For Haraway, recall, the figure of the cyborg is crucially both ontological *and* hermeneutical, both a material entity and a discursive construction, both a lived reality and an image ('A Cyborg Manifesto', p. 149). On the one hand, it names the real historico-material status of women living under late capitalism: 'the cyborg is our ontology' (p. 150). On the other hand, however, it is also the utopian signifier of a feminist and materialist politics of liberation to which we are summoned: the cyborg is an anti-foundational 'myth of political identity' (p. 173). If Haraway's decision to posit the cyborg as not simply an ontological condition but a strategic political metaphor enables her to extend the scope of her diagnosis to cover everything from domestic labour to science fiction, it also reveals an intriguing commitment to the performative force of the idea: it is not enough to simply *be* a cyborg to accomplish the work of political transformation because, as the famous conclusion to her essay makes clear, women must also *choose* to be a cyborg (as opposed to, say, a goddess). Such a critical stress on the moment of voluntarist self-recognition as the vehicle by which women begin to obtain liberation from what Haraway calls the 'informatics of domination' under late capital nonetheless betrays the extent to which, for all its impeccably left-deconstructionist credentials, her essay retains a deep investment in the idealism it seeks to critique. In Mark B. N. Hansen's otherwise sympathetic account, for instance, 'A Cyborg Manifesto' tells the story of an implicitly cognitive subject who is called to understand, gain agency over and strategically deploy their historico-material status as a cyborg in order to pursue a classically liberal project of emancipation through self-determination (*Embodying* Technesis, pp. 13–14).

For Katherine Hayles, critical posthumanism – and the state of embodiment with which, for her, it is synonymous – represent the most powerful philosophical antidote to the Cartesian idealism she associates with transhumanism but, once again, the ghost in the cyborg is not so easy to exorcise as one might think. It is made abundantly clear throughout the brilliant and richly textured narrative of *How We Became Posthuman* (1999) that Hayles' critique of the liberal subject does not amount to a wholesale anti-humanism: 'serious consideration', she insists, 'needs to be given to how certain characteristics associated with the liberal subject, especially agency and choice, can be articulated within a posthuman context' (*How We Became Posthuman*, p. 5). Accordingly, what 'the posthuman' seems to stand for in this context is less a literal state of Being-in-the-World – only 10 per cent of the US population are really 'cyborgs' in the technical sense of incorporating non-biological matter like prosthetic limbs and pacemakers into their bodies (p. 115) – so much as a new and more general act of subject-constitution capable of challenging the liberal paradigm. Take, for example, the claim that '[t]he defining characteristics [of the posthuman] involve the construction of subjectivity, not the presence of nonbiological components' (p. 4). To Hayles, like Haraway before her and, we will see, Hansen afterwards, posthumanism is not just an ontological condition which we embody, in other words, but an ethico-political task to which we must rise in order to do the cultural work necessary to transform late capitalist society:

> ...my dream is a version of the posthuman that embraces the possibilities of information technologies without being seduced by fantasies of unlimited power and disembodied immortality, that recognises and celebrates finitude as a condition of human being, and that understands human life is embedded in a material world of great complexity, one on which we depend for our continued survival. (p. 5)

Such a posthumanist 'dream' of embodiment in matter is, naturally, to be contrasted with the liberal humanist myth of mastery over nature that finds its consummation in transhumanism but, as Slavoj Žižek amongst others has noted, the philosophical territory the two occupy is not as far apart as it first appears.[9] Both offer narratives of emancipation and enlightenment from a prior state of tutelage; both tell the story of human life wilfully making and re-making its own quasi-mechanised nature in the absence of any pre-given natural or divinely ordained norm; both envisage the only *telos* of human existence to be the maximisation of its own welfare or survival.[10] If Hayles is right to insist that 'the posthuman need not be recuperated back into liberal humanism' (p. 287), in other words, it may only be because it never quite left that humanism behind in the first place. In fact, posthumanism's critique of autonomous liberal subjectivity in favour of an embodied and embedded collectivity is – provided one substitutes 'religion' for 'technology' as the name of that collective – entirely consonant with, say, the philosopher Charles Taylor's critique of what he calls the 'buffered self' from the vantage point of a Christian liberal communitarianism.[11]

Perhaps most importantly, though, Hayles' dream of a posthumanist future, just like the ironic myth that is Haraway's Cyborg Manifesto, reveals the extent of her own – both strategic and yet strangely metaphysical – idealism. Quite simply, posthumanism again takes the form of a consciousness-raising exercise whereby human beings are summoned by the new social imaginary of the posthuman (p. 6) to recognise themselves as the embodied beings that they always already are. What is going on here is not simply a bottom-up historical materialist critique of an obsolete superstructure, but a top-down transformation of a material condition by an idea. By becoming aware of the new possibilities for subject-formation that embodiment makes possible, readers may begin 'thinking of their actions in ways that have more in common with the posthuman than the human':

> Each person who thinks this way begins to envision herself and himself as a posthuman collectivity, an 'I' transformed into the 'we' of autonomous agents operating together to make a self. The infectious power of this way of thinking gives 'we' a performative dimension. People become posthuman because they think they are posthuman. (p. 6)

Why – if we are always already posthuman rather than liberal subjects – is it still so crucial to also *think* that we are? If Hayles certainly offers a potent counter-narrative to the humanist myth of an autonomous self, it has to be said that this performative thought experiment remains a curious way to set about deconstructing idealism: I think (or rather the multitude of sub-agents that make me up thinks) therefore I am posthuman. Presumably the allusion to Descartes' own '*Cogito ergo sum*' is intended to be ironic, but that does not change the fact that what is taking place here is an entirely Cartesian act of self-reflexivity in which a collective embodied *res cogitans* comes into existence through the act of thinking itself thinking. Just as Hayles prioritises the moment of *cogito me cogitare* as the true birth of the posthuman, so *How We Became Posthuman* continually privileges the *idea* of embodiment as the engine that drives its narrative of ever-extending cognition, agency and ultimately survival: '[t]o conceptualise the human in these terms is not to imperil human survival, but is precisely to enhance it' (*How We Became Posthuman*, p. 290). In the same way, Hayles' consistent assumption throughout *How We Became Posthuman* seems to be that if only we can only secure the right *idea* of the 'we' we call 'the human', humanity's position itself will be automatically transformed and it will regain – if not the total mastery of humanism – then at least a measure of agency over its own fate:

> Although some current versions of the posthuman point towards the anti-human and the apocalyptic, we can craft others that will be conducive to the long-range survival of human beings and of the other life-forms, biological and artificial, with whom we share the planet and ourselves. (p. 291)

In many ways, the work of new media theorist Mark B. N. Hansen offers the most compelling recent contribution to the field of critical posthumanism not simply because it offers a powerful corrective to what it diagnoses as the lingering idealism that afflicts contemporary continental philosophy of technology (from Heidegger up to and including Haraway and Hayles) but because it also furnishes one of the most philosophically sophisticated examples of the turn to the body that we currently possess.[12] It is only by building a strong theory of the body as the indispensable sensual and perceptual enframer of all media phenomena, Hansen insists, that continental philosophy can move beyond what he calls its persistent logic of '*technesis*': the reduction of the absolute material exteriority of technics to an abstract construction that exists only in relation to the *ego cogito* (*Embodying* Technesis, pp. 18–22). At the same time, though, Hansen's penetrating critique of the growing idealisation of technics in continental philosophy does not lead him to become an uncritical champion of the absolute exteriority of the technical real. Despite his strong desire to move beyond linguistic or cultural constructionism, Hansen emphatically rejects the kind of uncompromising technological anti-humanism of figures such as Friedrich Kittler who argue that – far from being dependent upon human beings – technogenesis is now taking place entirely *autonomously* of us, outstripping our human, all too human perceptual and affective ratios. Not only does Hansen rebut the claim that technical becoming has evolved to the position where it can now be said to have an artificial 'life' of its own – contra Kittler's determinism – but, as we will see, he continues to insist upon the embodied subject as the necessary site where technics takes on meaning (see, for example, *New Philosophy for New Media*, pp. 73–9). Yet, at times, such an insistence upon the anthropological ground of all technogenesis appears to leave behind its finely grained theoretical or empirical rationale and becomes something closer to an article of faith.[13] To give just one small but very telling example, Hansen offers a passionate critique of Jean-François Lyotard's essay 'Can Thought Go On Without a Body?' (*The Inhuman*, pp. 8–23) on the grounds that its history of human life from the vantage point of the death of the sun approximately 4.5 billion years from now reduces us to little more than pawns in a cosmological game of techno-evolutionary complexification:

> The unwavering privilege [Lyotard] grants the cosmic point of view is fundamentally disempowering, if not downright repugnant, given the urgency of our responsibility, as contemporary critics and as world citizens, to explore the *anthropology* of the technological age. In my opinion, moreover, Lyotard's striking nihilism is less the reflection of any actual or possible experiential reality than a result of his effort to ally his thinking rhetorically with the cosmological perspective he claims to be explicating. (*Embodying* Technesis, p. 67)[14]

Such a desire for philosophical anthropology over and against inhuman cosmology is, for all its claims to urgency, less an ontological or epistemological truth-claim than the expression of an ethico-political preference: it may well be an inconvenient truth for embodiment theory that the sun is dying – solar death is a form of technogenesis that will, after all, require no body to enframe it – but that does not mean it will not happen. If it is certainly correct to say that Lyotard's cosmological 'nihilism' disempowers the individual subject, this does not, of course, render it any less legitimate a perspective: a liberal humanist deeply invested in the idea of total mastery over nature could, after all, level exactly the same political objection of disempowerment at embodiment theory itself. Given the root-and-branch critique of social and linguistic constructionism that drives his account, Hansen's real problem with Lyotard's cosmology paradoxically seems to be that it is not constructionist *enough*: it is only when we begin to view the latter heuristically – as 'a cosmological fairy tale', no less, to compare with Haraway's 'Cyborg Manifesto' (p. 68) – that it regains the necessary embodied ground for technogenesis he deems to be essential. By strategically bracketing the ontological realism of Lyotard's cosmology in favour of a political preference, the theorist effectively boils its radical inhumanism back down to just another thought experiment that reveals 'what-it-means-to-be-human'. Just as Haraway sees embodiment less as an ontological given than an desideratum, so Hansen too necessarily transforms the body into a kind of necessary strategic fiction which affords us our last hermeneutical redoubt against the cosmological void of the real. Perhaps one might respond that there is nothing wrong with any of this – we will see in the next section that Hansen is quite right to detect a lingering idealism in Lyotard's cosmo-nihilism – but that does not change the fact that the former's position here is every bit as constructionist as the dominant techno-cultural paradigm he rejects. For me, we might even go so far as to speak of a logic of '*somesis*' at work here – a putting-into-discourse of the body to rival the textualisation of technology that is *technesis* – which puts the body into its now obligatory position within embodiment theory as the protagonist of a neo-humanist *Bildungsroman* of ever-increasing awareness, agency and survival:

> Only by preserving the value of our short-term human perspective will we retain the ability to invest our lives with significance – to intervene (to whatever extent possible) in struggles over the future of technology and to engage ourselves in the affirmative task of exploring new technologically mediated, posthuman forms of *human* agency. (p. 68, Hansen's emphasis)

What exactly is it about the body, in Hansen's view, that enables us to reclaim some measure of agency over the vicissitudes of blind technological becoming? To condense somewhat brutally a theory that gathers together and updates Benjamin (*Embodying* Technesis), Bergson (*New Philosophy for New Media*) and Merleau-Ponty (*Bodies in Code*), Hansen's consistent refrain is that embodiment is the necessary frame through which the abstract zeros and ones of digital code must pass in order to acquire meaning:

> The act of enframing information can be said to 'give body' to digital data – to transform something that is unframed, disembodied, and formless into concrete embodied information intrinsically imbued with (human) meaning. (*New Philosophy for New Media*, p. 31)

It is only because the body is what Bergson calls a 'centre of indetermination'[15] – a primary frame that selects, filters and gives form to the formless universe of images that circulate around it – that information gets constituted *as* information in the first place: 'this "originary" act of enframing information must be seen as the source of all technical frames (even if these appear to be primary), to the extent that these are designed to make information perceivable by the body, that is, to transform it into the form of the image' (*New Philosophy for New Media*, p. 11). At the same time, we must also bear in mind that this originary act of enframing is not cognitive but *affective*: Bergson makes clear that it is the body's sensorimotor capacity – its proprioceptive experience of itself as a source of possible action with particular needs and functions – that is the criterion by which it forms the relevant image from the universal flux (pp. 4–5). During *Bodies in Code* (which is the second volume in a proposed trilogy on new media) the theorist turns from Bergson to the later Merleau-Ponty's theory of the body-schema to fine-tune his conceptualisation of this pre-cognitive, operational and affective work of the body (*Bodies in Code*, p. 20). By constantly adapting itself in response to stimuli from its external environment, Merleau-Ponty's body-schema not only provides Hansen with a means of protecting his theory of embodiment against any charge of valorising a residual, pre-linguistic intuitionism but enables him to canvass – in a manner that at least superficially recalls Derrida and Stiegler's account of originary technicity – for the 'co-functioning of embodiment with technics' (p. 9). For Hansen, though, the philosophical upshot of embodiment remains largely the same throughout his readings of everything from high theory to new media art: contemporary media theory must (re-)turn to the body as the fleshly noesis of all informational noemata if it is to decisively break with both the representationalism that stretches from Descartes to post-structuralism, on the one hand, and the technological determinism that encompasses everyone from Moravec to Kittler, on the other. Now, despite (or because of) this attempt to posit a new absolute ground for meaning-making in the body, it is hard not to detect a very old desire at work here: Stephen Dougherty goes so far as to suspect 'a return to a firm basis, an originary ground from which we might build a body-oriented subjectivism to substitute for a debunked Cartesian subjectivism' in Hansen's philosophy.[16] Such a desire is arguably most manifest in Hansen's persistent stress on the auto-affection of the body – its sense of itself as the original form that has the capacity to give form to everything outside itself – as the necessary ground of all perception:

> The bottom line is that we are able to perceive images only because we sense ourselves as form. Perception, in short, depends on affectivity. What this means is that the auto-subjectivity of form-giving forms cannot be likened to a perception of self, but must be understood as a primary affectivity, a 'consciousness texture' that underlies and conditions all experience, including perceptual experience. (*New Philosophy for New Media*, p. 71)

From Raymond Ruyer's theory of the 'pure self-enjoyment' that constitutes the affective basis of all information (*New Philosophy for New Media*, p. 183) to Didier Anzieu's account of the '*infratactility*' – the self-acting or self-touching – that is the primary condition of all perception (*Bodies in Code*, p. 70), Hansen consistently seeks to locate a kind of philosophical ground zero in the possibility of a pure, embodied (if always technicised) auto-affection. If he is careful to distinguish this deep substratum of auto-affection from any cerebralist gesture of self-perception, it is worth noting that the 'body' of which Hansen speaks – which must first constitute itself in order to constitute the other – still performs an uncannily similar role to what Descartes calls 'mind': the idealist formula of 'I-think-myself-thinking' (*cogito-me-cogitare*) is seamlessly replaced by an embodied 'I sense-myself-sensing' or 'I touch-myself-touching'. Perhaps a sceptical reader might here recall Jacques Derrida's claim in *On Touching* that the haptic is – for all its pretensions to an unrivalled immediacy of intuition or contact whether it be in the act of self-touching or touching the other – no less metaphysical than the optic: the later Merleau-Ponty's phenomenology of the flesh, in particular, is somewhat harshly convicted by Derrida of presiding over a quasi-theological intuitionism, a metaphysical anthropology and (as we saw with Husserl in Chapter 5) a reduction of the originary technicity of touch to a spatio-temporal presence. In their rival accounts of Merleau-Ponty's body-schema, we encounter an unbridgeable caesura between Derrida and Hansen's concepts of the originary technicity of embodiment in spite of a certain superficial overlap of vocabulary: what Hansen portrays as the necessary embodied condition of what would otherwise be a purely abstract process of technogenesis, Derrida depicts as a gesture of transcendental mastery *over* the body's own irreducible technicity.[17]

2

In drawing this book to a close, however, I wish to briefly consider one final and somewhat less consoling vision of our posthumanist future: death. It is the very simple and yet radically unthinkable prospect of a world without human beings – a world in which the human race did not always exist and will all too soon cease to exist – that has begun to inspire a diverse group of recent philosophers, historians and scientists. As we mentioned at the beginning of this chapter, a new genre of philosophical futurology has emerged in recent years that seeks to narrate the story of the *anthropos* from the perspective of an (at least apparently) inhuman process of geological, biological, historical and physical complexification. For the later Jean-François Lyotard of *The Inhuman* and *Postmodern Fables*, recall, it is the absolute ontological limit imposed by the death of the sun in 4.5 billion years that forms the philosophical ground for his account of human life as merely a brief episode in a cosmic process of negentropy. Just as Lyotard's cosmological point of view compels us to think the history of the human otherwise, so a range of recent academic and popular accounts have also offered inhuman histories of human evolution: Manuel de Landa, Jared Diamond and others have offered what we might call materialist anti-anthropologies from the perspective of, variously, germs, plagues, guns and steel. Finally, the contemporary philosophical movement known as Speculative Realism has also invoked the trope of the inhuman as a means of rescuing the Absolute from the finitude of phenomenology, hermeneutics and cultural constructionism: Quentin Meillassoux's *After Finitude*, for example, grounds its critique of phenomenology's inescapable commitment to the subject-object correlation on the ontological ground zero imposed by a world without humans. To what extent, then, does the philosophy of inhumanism succeed in expunging the anthropocentric remainder that can still be found in transhumanism and posthumanism?

To be sure, the philosophy of the inhuman represents a decisive and (in my view) entirely welcome break with the anthropological machine that this book has adumbrated from Marx to Stiegler but it will be no surprise by now to learn of my suspicion that, even here, a certain spectral *anthropos* may still miraculously survive our total species extinction. Firstly, I think it is important to underscore that Lyotard's later work on *The Inhuman* is not simply a banal empirical discussion of artificial intelligence, environmental crises and the possibilities of post-organic life but a quasi-fabular extension of his critique of the Kantian sublime in works such as *The Differend* (1983).[18] It is Kant's claim in the 'Analytic of the Sublime', recall, that the sublime occurs when the inability of our sensible faculties to apprehend the power or magnitude of nature awakens reason's power to conceive of ideas (such as absolute power or magnitude) which transcend the sensible altogether.[19] As Yu Liu observes, the Kantian sublime recalls a kind of rational *fort/da* game whereby the subject turns its own inadequacy into an act of mastery: what is taking place here is nothing less than a 'fantasised vision of reason's total triumph at the expense of imagination'.[20] However, Lyotard's counter-claim is that there is an irresolvable differend between our sensible and rational faculties which ruins Kant's attempt to subsume the one under the other: the sublime consists precisely in the *unpresentability* of a sensible feeling in the form of a unifying concept like an Idea of Reason.[21] For the later Lyotard, 'the inhuman' is the site of precisely such a differend because, crudely speaking, it stages the incommensurable difference between an overpowering sensible feeling, on the one hand – life on earth is now obviously finite – and an infinite Idea of Reason – life will still continue to survive elsewhere – on the other:

> ...how to make thought without a body possible. A thought that continues to exist after the death of the human body. This is the price to be paid if the explosion is to be conceivable, if the death of the sun is to be a death like other deaths we know about. (*The Inhuman*, pp. 13–14)

If solar death's absolute limit upon human thought is itself to be thinkable, it must be brought into accord with an Idea of Reason and it is not difficult to see that the requisite idea here is the classic Kantian idea of 'the human' itself and particularly that of human progress.[22] Now, such an Idea of Reason, however necessary and unavoidable it may be, is consistently shown by Lyotard to be inadequate to the sublime feeling it seeks to present. By appealing to a supersensible idea of human survival, reason cannot trump the sensible feeling that life is coming to an end, because what that idea of the human makes possible is precisely a post-organic, post-affective, post-gendered and post-teleological life that will no longer bear any relation to anything we currently recognise *as* human. In James Williams' account, it is here that we encounter the unbridgeable differend between the sense of human finitude and the idea of human progress: 'the irony of Lyotard's thesis regarding the Idea of the human as a bet on a non-organic future is that the narrative that brings this idea to life for us is dehumanised' (*Lyotard and the Political*, p. 128).

It should be clear even from this brief account that – contra over-literal readings of his work – the later Lyotard is very far from being an apologist for the kind of grand cosmological narrative which simply demands that human beings evolve or die. As we have just seen, Lyotard rather seeks to expose such narratives as themselves dependent upon universalised Ideas of Reason like humanism that exhibit a fundamental disconnect with the irreducible particularities of the sensible world: 'Thought is inseparable from the phenomenological body' (*The Inhuman*, p. 23). Yet, as Williams goes on to note, Lyotard's commitment to the absolute incommensurability of the sensible and the rational faculties also imposes a limit on what his own thought can do: Lyotard himself can merely expose this moment of disconnect, not seek to suture the differences together in the form of a new, more secure judgement. Any 'politics of the differend' must restrict itself to the purely negative form of resisting the hegemony of an Idea of Reason over an irreducible plural set of sensible events if it is not to fall victim to the very appeal to a unifying higher reason that it seeks to critique in the Kantian sublime (*Lyotard and the Political*, pp. 116–18). For Lyotard, then, it would appear that Ideas of Reason – despite, or rather because of, their inability to overcome differences – paradoxically remain a necessary mediator between our sublime feeling and any given differend: we have no way of bearing witness to the absolute incommensurability of a genuine differend except through such a (failed) idea. If the later Lyotard remains reliant upon a minimal or negative idealism – because the sublime must always measure itself against an Idea of Reason that proves to be inadequate – it can also be accused of promulgating a distinctly pre-Kantian – indeed Burkean[23] – psychologism in its insistence upon *feeling* as the privileged signifier of that inadequacy. By mortgaging our ability to testify to the differend to nothing more than the contingency of a particular affect, Lyotard surrenders his ability to universally *prove* its incommensurability: Williams again points out that there is nothing necessary about such a feeling, nothing to guarantee its veracity, nothing even to show that it will occur more than once in one individual spectator (p. 116). In other words, the Lyotardian sublime cannot secure its own unique status as a testament to the failure of an Idea of Reason: Lyotard can very well say that a particular feeling cannot be presented by a particular idea at a particular moment but this is not the same thing as saying that it can *never* be presented, whether for myself or for others (pp. 116–17). Perhaps worst of all, Lyotard's commitment to the sublime cannot even prevent the minimal idealism of his philosophy of the inhuman from turning into a maximal one: we can find no necessary reason in essays like 'Can Thought Go On Without A Body?' – for all the ironies they expose – why the idea of post-organic life should not eventually triumph over the sensible feeling that life on earth is coming to an end.

For philosopher historians such as Manuel De Landa, of course, it is a Deleuzean neo-materialism – rather than the Kantian sublime – that furnishes the philosophical means to think the inhuman. It is symptomatic that De Landa's first book, *War in the Age of Intelligent Machines*, is written from the perspective of a hypothetical future robot historian dedicated to uncovering the historical origins of his own species, because his entire body of work can be read as just such an exercise in *prosopopoeia*.[24] As anyone familiar with his philosophical itinerary will be aware, each subsequent project takes an over-determined anthropological field – whether it be warfare, historiography or the social – and re-narrates it from the radically de-familiarising perspective of non-human actors: geology, economics or technology.[25] To clear the ground for such an exercise, all vestiges of anthropocentrism – whether it be the empirical privilege granted to the human observer as final arbiter over the real or the teleological assumption of linear human progress – must be purged and the mythical *anthropos* plunged back into the dynamic, self-organising and constantly adapting flow of matter and energy from which it emerged:

> [O]ur individual bodies and minds are mere coagulations or decelerations in the flows of biomass, genes, memes, and norms. (*A Thousand Years of Non-Linear History*, p. 258)

However, for all the force of such attempts to ventriloquise matter, there are times when we can see the author's lips moving. For De Landa, in a manner reminiscent of Haraway and Hayles, matter-energy is not simply an ontological given to be reckoned with but a norm that must be realised: the human species is persistently called upon to take a more 'experimental attitude to reality' and to show an 'increased awareness' of the self-organisation inherent in even the humblest phenomena (pp. 273–4). Such a demand accords a limited, but in this context still surprising, degree of privilege to the thinking subject as agent and cause of material complexification: 'Changing our way of thinking about the world is a necessary first step, but it is by no means sufficient: we will need *to destratify reality itself*' (p. 274). Perhaps – at least in some of his more rhetorical flights – De Landa's valorisation of matter even begins to resemble a quasi-Romantic natural piety: we are told that neo-materialism contains 'a new kind of hope', for instance, and 'a positive, even joyful conception of reality' (p. 274). If it is certainly gratifying to learn that 'many of the most beautiful and inspiring things on our planet have been created through destratification' (p. 274), the fact that De Landa feels compelled to reassure us on this score merely adds to the suspicion that there is something more than a pure ontological realism going on here. Who cares, frankly, whether neo-materialism contains hope, beauty and joyfulness – if not the very anthropological observer that De Landa is seeking to subtract? In the same way, the formal conceit of the robot historian does not do away with this anthropological observer so much as devolve it onto matter itself: what stubbornly remains in this audacious attempt to do away with anthropocentrism is once again a quasi-anthropomorphic 'point of view' (*War in the Age of*

Intelligent Machines, p. 10) from which the process of technogenesis may be witnessed.

Finally, and very briefly, I also want to mention the emergence of what is arguably the most remarkable recent attempt to think an inhuman world: Quentin Meillassoux's Speculative Realism. It is what Meillassoux calls 'correlationism' – the anthropocentric assumption that the external world is merely the correlate of the subject that conceives of it – that is the target of his *After Finitude: An Essay on the Necessity of Contingency.* According to Meillassoux, the story of modern philosophy (from the Kantian transcendental subject, through the noetico-nomatic relation of Husserlian phenomenology, the mutual belonging together of Being and beings articulated by Heidegger's *das Ereignis*, all the way to the quasi-theological ethics of Levinas or the later Derrida) is the story of the reduction of the Absolute to something that is dependent upon the conceptual, linguistic or cultural representation of a thinking subject. However, *After Finitude*'s thesis is that such correlationism is fundamentally incoherent because it is unable to make sense of the observer-independent reality revealed by physics – a universe that existed before humans were there to conceive of it (what the philosopher technically terms 'ancestrality') and that will continue to exist after our extinction. To simplify a little brutally, Meillassoux argues that, when taken to its logical conclusion, correlationism effectively finds itself in a double-bind: either it must deny the existence of the external world altogether – and succumb to the kind of absurd absolute idealism that renders humanity immortal – or it must admit that its epistemological subjectivism contains an implicitly objective claim about what the philosopher calls 'the Great Outdoors [*le Grand Dehors*]'. Such an implicit ontological claim can, for example, be glimpsed in Hume's famous account of the problem of causality in the *Enquiry Concerning Human Understanding*: what the empirical philosopher depicts as a mere epistemological limit upon thought – we cannot rationally know that there is a necessary link between cause and effect – is actually a misrecognised insight into the real, if chaotic, properties of the universe itself. For Meillassoux, in other words, Hume's misplaced insight opens the door for a singularly audacious ontological claim, namely, that an absolute, observer-independent reality *exists* but that it is not subject to any principle of reason and therefore absolutely *contingent*:

> Everything could actually collapse: from trees to stars, from stars to laws, from physical laws to logical laws; and this not by virtue of some superior law whereby everything is destined to perish, but by

> the virtue of the absence of any superior law, capable of preserving anything, no matter what, from perishing. (*After Finitude*, p. 53)

If any serious assessment of Meillassoux's thesis – from its dismissal of the entirety of post-Kantian philosophy as a species of correlationism to its claim that phenomenology's abandonment of the Absolute leads it to preside over a disturbing 'becoming-religious of thought' (p. 46) – obviously lies well beyond the scope of this book, a great deal of its critical reception thus far has understandably focused on the question of whether *After Finitude* overcomes the idealism it condemns. Perhaps the fact that the bulk of the book consists in offering an immanent critique of correlationism – which seeks to show how, even when taken on its own terms, it does not work – means that it risks conceding too much ground to the enemy: it seems, at times, as if the empire of thought is capable of being abolished by thought alone. In many ways, though, we can safely say that Meillassoux's essay returns us to the classic antinomies which the problem of originary technicity throws up – man and matter, idealism and materialism, anthropocentrism and techno-scientific realism – and it is a tribute to his work that it raises as many questions as it answers. To what extent, for example, does *After Finitude*'s investment in a pure mathematical ontology as the privileged means of access to the real – even if is Cantor's claim that there is an open, unending series of ever-larger infinite numbers – risk contradicting the argument that there are *no* a priori principles, no sufficient reason whatsoever, which governs or explains that contingent reality? How far can Cantor's Non-All be taken out of the domain of pure mathematics and straightforwardly applied to a concrete situation in our own universe with an (at least apparently) closed set of possibilities? Does the absolute contingency Meillassoux discovers in the physical sciences have anything positive to say anything about the relative 'stability' of cause and effect as we currently observe it in, say, evolutionary biology? Can we detect an undisclosed ethico-political as well as an ontological dimension to Meillassoux's critique of correlationism – particularly in the claim that it permits religion entry via the backdoor – that needs to be acknowledged and justified?

3

This book has sought to offer a critical genealogy of what has, for better and worse, come to be known as 'originary technicity'. Its guiding hypothesis has been that continental philosophy of technology from

Marx onwards has consistently embraced a generalised technicity as the empirico-transcendental condition not only of human life but also of life more generally. As I have tried to show, for all the real and many differences between them, Marx and Engel's theory of labour; Freud and Lacan's materialist definition of consciousness; Heidegger's ontological reading of technology; Derrida's articulation of originary technicity; Stiegler's account of epiphylogenesis; and Haraway, Hayles and Hansen's theories of the posthumanism articulate a powerful collective logic. For each thinker, what is nominally posited as *outside* the sphere of the living is folded back *inside* life as its enabling condition.

Yet, I have also repeatedly argued (perhaps excessively so for some readers) that such a radical act of expropriation is too often compromised by an, equally machinic, gesture of re-appropriation. It has been my larger contention that the organico-inorganic fold of originary technicity is *itself* re-folded so that it may become the ground for a new, albeit aporetic, anthropology. Such, I have suggested, turns out to be the ironic fate of a great deal of continental philosophy of technology: what should turn the *anthropos* inside out – by showing how, even in its very self-referential ipseity, it is constitutively open to its factical, inorganic outside – is itself internalised, anthropologised, humanised. In this respect, I would concur with Timothy Clark's claim that originary technicity frequently ends up producing little more than a non-essentialist anthropology ('Derrida and Technology', p. 247).

To be sure, this anthropological machine takes many forms – from its occlusion of any human-independent form of technogenesis to its reduction of technics to a material exemplum of some prior and more abstract philosophical problematic – but arguably the most persistent and violent we have observed is its re-introduction of a spectral anthropomorphic subject or witness to act as the presiding judge, so to speak, in the case of life versus technics. Such is the (witting or unwitting) position occupied by, for example, Marx's collective labouring subject who is able to liberate himself from technological subsumption; Lacan's human observer who returns to pass judgement upon the materialist state of consciousness; Heidegger's poet or questioner who can piously ask after the ontological essence of technology; Derrida's philosopher who can deconstruct the metaphysics of time and space at work in Real Time technologies; Stiegler's spectator who is able to consume/produce digital media critically rather than passively; and, finally, Lyotard's dying sun and de Landa's robot historian from whose future anterior point of view the cosmological process of technogenesis will have taken place.

Perhaps, then, the real story of the theory of originary technicity – from historical materialism to psychoanalysis, from fundamental ontology to deconstruction, from libidinal materialism to the different species of critical posthumanism and embodiment theory – is the reduction of what I have called the empirico-transcendental condition *of* thought to just one more idea *to be* thought – and thought, moreover, by an all too familiar thinking thing (*sum res cogitans*). If we have identified some important exceptions to this rule throughout this book (from the kind of structural coupling or transduction of human and tool articulated in Marx, Leroi-Gouhran, Derrida and Stiegler; through the techno-material complexification described by Lyotard; the return to the body in the work of critical posthumanists like Haraway and Hansen, up to Meillassoux's mathematical ontology) it has to be said that they remain largely that – exceptions – which, for all their rejection of anthropologism, cannot entirely guarantee that they, too, will not, in the end, arrive accompanied by the silent 'I think' of the *cogito*.

For all its claims to be 'always already there' – all the way back to the first cell – it is thus tempting to conclude this book by saying that originary technicity has still not quite *begun*. It is easier, of course, to ask questions than to provide answers and this study has remained largely on the level of critique. Yet it is also a modest invitation to take up the challenge of thinking originary technicity otherwise. Perhaps (and here is the slightly perverse speculation with which I would like to conclude) the ontological price to be paid for any philosophy of technology that does not wish to reduce technics to a mere toy of thought may ultimately be the surrender of *technics* – both the word and the idea – itself. If what passes under the name of 'technics' is indeed the empirico-transcendental condition of thought, of consciousness, of temporalisation and so on; if technics is inscribed in everything from the flesh and blood of the human hand to the zeros and ones of information; if technics is, so to speak, 'everywhere', then it also follows that technics is *nowhere, nothing* – technics 'is' *not*. From the early Derrida's claim that 'a new name must perhaps be found in order to remove [technology] from its traditional problematic' ('Freud and the Scene of Writing', p. 228) to Stiegler's argument that technics can never remain on one side of any evaluative hierarchy (*Technics and Time 1*, p. 14), continental philosophy of technology has, in fact, often manifested what we might call a kind of death drive which wills its principal *raison d'être* out of existence. In the recent work of Jean-Luc Nancy, for example, it is clear that any attempt to ontologise technicity as the basic state or condition of existence must inevitably lead to its disappearance as a regional ontology: Nancy is happy to admit that, because existence just *is* technological through and through, then 'there *is* no technology as "such" '.[26] What if the future of originary technicity is less the death of the human, consciousness or life than of the thing we still stubbornly, ideally and humanly persist in calling 'technology'?

Notes

1 Life

1. Jacques Derrida, 'Nietzsche and the Machine' trans. by Richard Beardsworth, in *Negotiations: Interventions and Interviews 1971–2000* ed. by Elizabeth Rottenberg (Stanford, CA: Stanford University Press, 2002), p. 248. All further references will be given in the text. In this book, I follow recent precedent in adopting the term 'technics' (*la technique*), as opposed to the more classical 'technology', to name the radical and generalised concept of technicity at work in contemporary philosophy.
2. François Jacob, *La logique du vivant: Une histoire de l'héredité* (Paris: Gallimard, 1970).
3. Jacques Monod, *Chance and Necessity, On the Natural Philosophy of Modern Biology* (Harmondsworth: Penguin, [1970] (1997), p. 45. See also Richard Dawkins, *Climbing Mount Improbable* (London: Penguin, 1996) and Susan Blackmore, *The Meme Machine* (Oxford: Oxford University Press, 2000) for different applications of the machine metaphor for molecular and evolutionary biology.
4. See Norbert Weiner, *Cybernetics or Control and Communication in the Animal and the Machine* (Cambridge, MA: MIT Press, 1948) amongst many other foundational texts on cybernetics.
5. Jean-François Lyotard, *The Inhuman: Reflections on Time* trans. by Geoffrey Bennington and Rachel Bowlby (Cambridge: Polity, 1991), p. 52. 'As anthropologists and biologists admit', Lyotard writes, 'even the simplest life forms, infusoria (tiny algae synthesised by light at the edge of tidepools a few million years ago) are already technical devices. Any material system is technological if it filters information useful to its survival, if it memorises and processes that information and makes inferences based on the regulating effect of behaviour, that is, if it intervenes on and impacts its environment so as to assure its perpetuation at least' (p. 12). All further references will be given in the text.
6. Jacques Lacan, *The Seminar of Jacques Lacan, Book I: Freud on Technique* trans. by John Forrester and ed. by Jacques-Alain Miller (Cambridge: Cambridge University Press, 1988) and *The Seminar of Jacques Lacan, Book II: The Ego in Freud's Theory and in the Technique of Psychoanalysis 1954–5* trans. by Sylvana Tomaselli and ed. by Jacques-Alain Miller (Cambridge: Cambridge University Press, 1988). All further references will be given in the text.
7. Gilles Deleuze and Felix Guattari, *A Thousand Plateaus: Capitalism & Schizophrenia* trans. by Brian Massumi (Minneapolis, MN: University of Minnesota Press, 1987), p. 406. In Deleuze and Guattari's account, the machinic phylum may be defined as 'the flow of matter in continuous variation, conveying singularities'.
8. Michel Foucault, *Birth of Biopolitics: Lectures at the Collège de France 1979–80* trans. by Graham Burchell (Basingstoke and New York: Palgrave Macmillan,

2008); Giorgio Agamben, *Homo Sacer: Sovereign Power and Bare Life* trans. by Daniel Heller-Roazen (Stanford, CA: Stanford University Press, 1998). See Michael Dillon and Julian Reid, *The Liberal Way of War: Killing to Make Life Live* (London: Routledge, 2008) for an acute analysis of the politicisation of both molecular biology and information theory in contemporary international relations. All further references will be given in the text.

9. Friedrich Kittler, *Discourse Networks 1800–1900* trans. by Michael Metteer with Chris Collins (Stanford, CA: Stanford University Press, 1995); Bernard Stiegler, *Technics and Time 1: The Fault of Epimetheus* trans. by Richard Beardsworth and George Collins (Stanford, CA: Stanford University Press, 1998) and Jean-Luc Nancy, *Corpus* (Paris: Editions Métailié, 2000) amongst many other texts. All further references will be given in the text.
10. David Wills, *Dorsality: Thinking Back Through Technology and Politics* (Minneapolis, MN and London: University of Minnesota Press, 2008), p. 3. All further references will be given in the text.
11. Plato, *'Protagoras' and 'Meno'* trans. by Robert C. Bartlett (Ithaca, NY: Cornell University Press, 2004), 81c–d. In Plato's words, 'since the soul is immortal and has come to be many times and has seen both the things here and those in Hades – in fact all things – there isn't anything it hasn't learned. As a result, its being able to recollect what pertains to virtue and other things is nothing to be wondered at, since it also knew them all previously. For, nature as a whole being akin and the soul having learned all things, nothing prevents someone, once he has recollected just one thing – what human beings call "learning" – to discover all else.' All further references will be given in the text.
12. Jacques Derrida, 'Plato's Pharmacy', in *Dissemination* trans. by Barbara Johnson (Chicago, IL: University of Chicago Press, 1982), pp. 67–186, 111. All further references will be given in the text.
13. Bernard Stiegler, '*Anamnēsis* and *Hypomnēsis*: The Memories of Desire' trans. by Arthur Bradley and François-Xavier Gleyzon, in *Technicity* ed. by Arthur Bradley and Louis Armand (Prague: Charles University Press, 2007), pp. 15–41, 27. All further references will be given in the text.
14. To begin with, Plato's own philosophy is far too complex to be reduced to a monolithic idealism that prioritises theoretical knowledge of the Forms (*epistemē*) as an end in itself over purely technical craft knowledge or know-how (*technē*). It is clear that – contrary to Bernard Stiegler's suggestion (*The Fault of Epimetheus*, p. 1) – *epistemē* and *technē* overlap throughout the Dialogues almost to the point of becoming interchangeable. On the one hand, *technē* is not simply a craft but a form of knowledge: the physician's craft, for example, is the *knowledge* of health (*Charmides*, 165c). On the other hand, *epistemē* is not simply theoretical knowledge of the Forms but of the *craft* of imitating them in the finite world (*Republic*, 484c–d). In the *Timaeus*, we even find a kind of Platonic 'originary technicity': the Demiurge or divine artisan crafts the Universe by imitating the world of Forms (*Timaeus*, 28a). See Anne Balansard, *Techné dans les dialogues de Platon* (Sankt Augustin: Academia Verlag, 2001).
15. Aristotle, *Physics* trans. by Robin Waterfield (Oxford: World's Classics, 1999), 192b–193b. In Book 2 of the *Physics*, Aristotle writes: 'Of things that exist, some exist by nature, some from other causes. "By nature" the animals

and their parts exist, and the plants and the simple bodies (earth, fire, air, water) – for we say that these and the like exist "by nature". All the things mentioned present a feature in which they differ from things which are not constituted by nature. Each of them has within itself a principle of motion and of stationariness (in respect of place, or of growth and decrease, or by way of alteration). On the other hand, a bed and a coat and anything else of that sort, *qua* receiving these designations i.e. in so far as they are products of art – have no innate impulse to change.' All further references will be given in the text.

16. Martin Heidegger, 'On the Essence and Concept of *φύσις* in Aristotle's *Physics* B, I' trans. by Thomas Sheehan, in *Pathmarks* ed. by William McNeil (Cambridge: Cambridge University Press, 1998), pp. 183–230, see especially pp. 195–6. In Heidegger's account, *phusis* has been reduced by modern biology into a kind of artefact with a built-in motor that enables it to make itself. All further references will be given in the text.
17. Aristotle, *Politics* trans. by T. A. Sinclair and rev. by Trevor J. Saunders (Harmondsworth: Penguin, 1981), Book 1, Chapter 4, 1253b. In many ways, the Aristotelian metaphysical hierarchy of theory over practice is paralleled by the political hierarchy of freedom over enslavement.
18. Amos Funkenstein, *Theology and the Scientific Imagination from the Middle Ages to the Seventeenth Century* (Princeton, NJ: Princeton University Press, 1986), pp. 28–42.
19. Georges Canguilhem, *La Connaissance de la vie* (Paris: Vrin, 1985). All further references will be given in the text. In this analysis, Canguilhem draws on Alfred Espinas' comparative study of *De Motu Animalium* and *Quaestiones Mechanicae*. See Alfred Espinas, 'L'Organisation ou la machine vivante en Grèce au IVe siècle avant J-C', *Revue du métaphysique et de morale* (1903), pp. 702–15.
20. René Descartes, *Meditations on First Philosophy* trans. by John Cottingham (Cambridge: Cambridge University Press, 1986), p. 58. All further references will be given in the text. In the *Principles of Philosophy*, Descartes explicitly refuses the Aristotelian opposition between *phusis* and *technē*: the philosopher sees no essential difference between the 'clock constructed with this or that set of wheels' and a 'tree which grew from this or that seed'. See René Descartes, *The Philosophical Writings of Descartes* trans. by John Cottingham Robert Stoothoff and Dugald Murdoch, volume 1 (Cambridge: Cambridge University press, 1985), p. 288.
21. Julien de la Mettrie, *Man a Machine* trans. by G. C. Bussey (Chicago, IL: Open Court, 1912), p. 128.
22. Immanuel Kant, *Critique of Judgement* trans. and ed. by Werner S. Pluhar (Indianapolis, IN: Hackett, 1987). All further references will be given in the text.
23. Pheng Cheah, 'Human Freedom and the Technic of Nature', *Differences: A Journal of Feminist Cultural Studies* 14: 2 (2003), pp. 1–26. In Cheah's account, 'the technic of nature is nothing other than the process of sending and giving that renders nature vulnerable to supplementation' (p. 21).
24. Anson Rabinbach, *The Human Motor: Energy, Fatigue and the Origins of Modernity* (Berkeley, CA: University of California Press, 1992), pp. 72–80. All further references will be given in the text.

25. Giorgio Agamben, *The Open: Man and Animal* trans. by Kevin Attell (Stanford, CA: Stanford University Press, 2005), pp. 26, 33–8. All further references will be given in the text.
26. Gilbert Simondon, *Du mode d'existence des objets techniques* (Paris: Aubier, 1958), pp. 19–20 and *L'individuation psychique et collective* (Paris: Aubier, 1989).
27. André Leroi-Gouhran, *Gesture and Speech* trans. by A. Bostock Berger (Cambridge, MA: MIT Press, 1993), p. 42. All further references will be given in the text.
28. Jacques Derrida, *Of Grammatology* trans. by Gaytri Chakravorty Spivak (Baltimore, MD and London: Johns Hopkins University Press, 1976), p. 9 and *passim*. All further references will be given in the text.
29. See, for example, Christopher Johnson, *System and Writing in the Philosophy of Jacques Derrida* (Cambridge: Cambridge University Press, 1993); Arkady Plotnitsky, *Complementarity: Anti-Epistemology After Bohr and Derrida* (Durham, NC: Duke University Press, 1994); and, most recently, Cary Wolfe, 'Meaning and Event; or, Systems Theory and "The Reconstruction of Deconstruction" ', in *What is Posthumanism?* (Minneapolis, MN and London: University of Minnesota Press, 2010), pp. 3–30. All further references will be given in the text.
30. Humberto R. Maturana and Francisco J. Varela, *Autopoiesis and Cognition: The Realisation of the Living* (Dordrecht: D. Reidel, 1980). All further references will be given in the text.
31. Humberto R. Maturana and Francisco J. Varela, *The Tree of Knowledge* (Boston, MA: New Science Library: 1987). In Valera's later work on emergence, biosystems operate through a complex feedback loop of input and output whereby each new input from the environment introduces minor deviations into the system. See *Toward a Practice of Autonomous Systems: Proceedings of the First European Conference on Artificial Life* ed. by Francisco Valera and Paul Bourgine (Cambridge, MA: MIT Press. 1992).
32. For many critics, as we will see, continental theories of originary technicity are guilty of either excessive idealism or scientific positivism: technics is either raised to the level of pure abstraction as if were merely a set of symbols or reduced to a simple empirical object as if it were synonymous with the latest inventions. It is my hypothesis in this book that originary technicity actually performs, and requires, a radical re-organisation of the empirico-transcendental opposition. On the one hand, it is a Kantian transcendental condition for experience that, nevertheless, remains part of a finite, empirical environment. On the other hand, it is an empirical environment that, nevertheless, transcendentally exceeds all reduction to a mere regional object or *positum*. In this sense, it is of a piece with the continental philosophical re-working of the empirico-transcendental taking place under such concepts as Foucault's historical a priori; Deleuze's transcendental empiricism and Derrida's quasi-transcendental.
33. In *Embodying* Technesis*: Technology Beyond Writing* (Ann Arbor, MI: University of Michigan Press, 2000) and a number of subsequent works, Mark B. N. Hansen attacks what he calls the logic of '*technesis*' – the putting-into-discourse of technology – at work in continental philosophy of technics which reduces the absolute material exteriority of technics vis-à-vis the

Cartesian subject to an abstract, linguistic and phenomenological exteriority that exists only in relation to the *ego cogito*. For me, this thesis is a powerful one – and what follows is, both positively and negatively, influenced by it – but my own claim is subtly different: I seek to target a more general (conscious or unconscious) philosophical anthropocentrism – which is common to *both* post-Heideggerian phenomenology *and* Hansen's own embodiment theory – that effectively anthropomorphises the question of technics. All further references to *Embodying* Technesis will be given in the text.

34. Timothy Clark, 'Deconstruction and Technology', in *Deconstructions: A User's Guide* ed. by Nicholas Royle (Basingstoke: Palgrave Macmillan, 2000), pp. 238–57, 247. In fairness to Clark, he is referring to the specific interpretation of Derrida's philosophy of technics advanced by Richard Beardsworth rather than to the work itself.

2 Labour

1. Karl Marx, *Capital*, volume 1 trans. by Ben Fowkes (Harmondsworth: Penguin, 1990), p. 998. All further references will be given in the text.
2. Kostas Axelos, *Marx, penseur de la technique: De l'aliénation de l'homme à la conquête du monde* (Paris: Editions de Minuit, 1967). According to Derrida, Marx is 'one of the first thinkers of technics', *Specters of Marx: The State of the Debt, The Work of Mourning and the New International* trans. by Peggy Kamuf (New York: Routledge, 1994), p. 170. In *La Technique et le temps 3: Le temps du cinéma et la question du mal-être* (Paris: Galilée, 2003), Bernard Stiegler echoes this view: Marx is 'above all the one who, for the first time, thinks technics philosophically', pp. 135–6n. Translation mine. All further references will be given in the text.
3. Etienne Balibar, *La Philosophie de Marx* (Paris: La Découverte, 1993), p. 3.
4. Alfred Schmidt, *The Concept of Nature in Marx* trans. by Ben Fowkes (London: New Left Books, 1970). All further references will be given in the text.
5. Louis Althusser and Etienne Balibar, *Reading Capital* trans. by Ben Brewster (New York: New Left Books, 1970), p. 171.
6. Anson Rabinbach, *The Human Motor: Energy, Fatigue and the Origins of Modernity* (Berkeley, CA: University of California Press, 1992), pp. 72–80 and Amy E. Wendling, *Marx on Technology and Alienation* (New York: Palgrave Macmillan, 2009). All further references will be given in the text.
7. Jacques Derrida, 'Politics and Friendship: An Interview with Jacques Derrida' trans. by Robert Harvey, in E. Ann Kaplan and Michael Sprinker eds *The Althusserian Legacy* (London: Verso, 1993), pp. 183–231, 191.
8. Karl Marx, *The Economic and Philosophic Manuscripts of 1844* (London: Lawrence & Wishart, 1973), p. 113. All further references will be given in the text.
9. G. W. F. Hegel, *Hegel's Philosophy of Nature* trans. by M. J. Perry (London: Allan & Unwin, 1970), p. 204.
10. To be sure, the later Marx still dogmatically maintains that human beings are the only beings who labour: 'we pre-suppose labour in which it is an exclusively human characteristic' (*Capital*, p. 284). As this line of argument implies, Marx falls back on an – Aristotelian – distinction between human

and animal labour: what distinguishes the architect from the bee is not that one labours and the other does not – if anything the bee is *more* productive than the architect – but rather that human labour is the hylomorphic imprint of form upon matter. For Marx, the genius of the human architect is not simply that he works upon nature according to his immediate needs – like the bee – but rather that he is able to form an *idea* of those needs in advance and project their realisation in material form: 'At the end of every labour process, a result emerges which had already been conceived by the worker [*der Vorstellung des Arbeiters*] at the beginning, hence already existing ideally' (p. 284; see also the *1844 Manuscripts*, pp. 113–14). In the theory of labour advanced in *Capital*, however, Marx inverts Aristotelian hylomorphism: it is the matter that precedes the idea rather than the other way around.

11. Andy Clark, *Natural-Born Cyborgs: Minds, Technologies and the Future of Human Intelligence* (New York: Oxford University Press, 2003). All further references will be given in the text.
12. It is difficult to locate any precise philosophical or scientific antecedent for the kind of exchange of forces described by Marx in this passage. According to Schmidt, Marx obtains the theory of a metabolic exchange of matter or substance between man and nature from Jacob Moleschott (*The Concept of Nature in Marx*, pp. 87–93). For Rabinbach, however, what is taking place is less a metabolic exchange of matter than a thermodynamic conversion of energy between two bodies – man and nature – that are qualitatively equivalent: 'labour is no longer a creative or singularly human act' (*The Human Motor*, p. 77). In Amy Wendling's recent account, though, Marx never fully signs up to the thermodynamic paradigm: 'Marx refuses to assimilate human labour fully to the model of work done by an animal or machine' (*Marx on Technology and Alienation*, p. 81).
13. Karl Marx, *Grundrisse: Foundations of the Critique of Political Economy* trans. by Martin Nicolaus (London: Penguin, 1973), p. 494. All further references will be given in the text.
14. Karl Marx and Friedrich Engels, *The German Ideology* (London: Lawrence & Wishart, 1970), §1, p. 49. All further references will be given in the text.
15. Friedrich Engels, 'The Part Played by Labour in the Transition from Ape to Man', in *The Dialectics of Nature* trans. by Clemens Dutt (London: Lawrence & Wishart, 1940), pp. 270–97. All further references will be given in the text.
16. At the time Engels wrote his essay, the fossil record was very sparse, comprising little more than a few fragments of jawbones. In many ways, it was not until the 1959 discovery of Zinjanthropus that the fossil record began to catch up with Engels' own theory. See Charles Woolfson, *The Labour Theory of Culture: A Re-Examination of Engels' Theory of Human Origins* (London: Routledge and Kegan Paul, 1982).
17. Yet, as we saw with Marx, there is a tension in Engels' text around the question of human and animal labour. On the one hand, he, too, clearly wishes to stamp the human being as the *only* being who truly labours: 'The labour process begins with the making of tools', he says, and 'no simian hand has ever fashioned the crudest stone knife' (*Dialectics of Nature*, pp. 286, 281). On the other hand, as we will see, he also makes clear that the human being

is the 'creation' of (presumably prehominid) labour and the human hand the 'product' of the (again logically prehominid) labour process. In other words, the human being is – impossibly – posited as both *cause* and *effect* of labour.

18. On the one hand, *The Poverty of Philosophy* asserts that technological production determines needs: 'Most often needs arise directly from production or from a state of affairs based on production', *Selected Writings* ed. by D. McLellan (Oxford: Oxford University Press, 1977), p. 42. On the other hand, the *Grundrisse* identifies a complex of factors that drive machine-stage capital: ' dissection . . . through the division of labour . . . gradually transforms the workers' operations into more and more mechanical ones, so that at a certain point a mechanism can step into their places' (p. 704). See Bruce Bimber, 'Three Faces of Technological Determinism', in *Does Technology Drive History?* ed. by M. R. Smith and L. Marx (Cambridge, MA: MIT Press, 1994), pp. 79–100.
19. Keith Ansell Pearson, *Viroid Life: Perspectives on the Transhuman Condition* (London: Routledge, 1997). In Ansell Pearson's view, technological determinism produces a 'completely reified grand narrative of technology as the true agent and *telos* of natural and (in)human history, but also restricts technics to *anthropos*, binding history to anthropocentrism' (p. 124).
20. Bernard Stiegler, *Economie de l'hypermateriel et psychopouvoir: Entretiens avec Philippe Petit et Vincent Bontems* (Paris: Mille et une nuits, 2006), p. 54. In many ways, Stiegler's own recent critique of the epoch of 'hyperindustrialisation' – where the industrialisation of production is superseded by an industrialisation of consumption which destroys all capacity for psychic individuation – might be read as just such a revisionary reading of Marx.
21. Perhaps Marx never goes so far as to assert the *total* autonomy of the machine – even the steam engine requires regulation and maintenance – but Louis Armand nevertheless argues that the Marxian machine comes close to what Maturana and Valera later call autopoiesis: 'the capacity of a machine to transform and reproduce itself'. See Louis Armand, 'Technics and Humanism', in *Event States: Discourse, Time, Mediality* (Prague: Charles University Press, 2007), pp. 225–61.
22. If Marx also insists upon a distinction between objectification (*Vergegenständlichung*) – as the basic condition of human species-existence – and alienation (*Entäusserung*) – as the historic form objectification takes under capitalism – what I wish to stress here is the continuity between the two: it is precisely the objectification of knowledge, memory and desire in the first place that makes them ripe for capitalist exploitation. See Karl Marx, *Grundrisse*, pp. 831–2.
23. Bernard Stiegler, '*Anamnēsis* & *Hypomnēsis*: The Memories of Desire', p. 18. In making this point, I am obviously departing from Stiegler's own view that Marx 'does not yet see human life as ex-istence' inasmuch as ex-istence is inextricably bound up with mnemo-technical supplements (p. 22): my claim is that Marx's theory of labour *does* posit life as ex-istence even if it does not always cash out this insight politically.
24. Jennifer Bajorek, 'Animadversions: *Tehkne* after Capital/Life after Work', *Diacritics* 33: 1 (2003), pp. 42–59, 58.
25. Peter Osborne, 'Marx and the Philosophy of Time', *Radical Philosophy* 147 (2008), pp. 15–22. All further references will be given in the text.

26. Moishe Postone, *Time, Labour and Social Domination: A Reinterpretation of Marx's Critical Theory* (Cambridge: Cambridge University Press, 1993), pp. 200–16. In Postone's account, clock-time is capital time par excellence: the increasing abstraction, quantification and spatialisation of time from the Middle Ages onwards goes hand in hand with the increasing capitalisation, objectification and commodification of labour.
27. For Stiegler, it is only via the mnemo-technical preservation of a past that it has not lived that *Dasein* is able to grasp its finitude: 'How does *Dasein*, on the basis of its essential temporality, gain access to its historiality? How is this historiality essential to its temporality, if not through a durable, necessarily pros-thetic fixing of the past, or rather of the "before" as what has happened [*comme ce qui s'est passé*]?', *The Fault of Epimetheus*, p. 236.
28. Antonio Negri, *Marx Beyond Marx: Lessons on the* Grundrisse trans. by Harry Cleaver, Michael Ryan and Maurizzio Viano (New York: Autonomedia/Pluto, 1991).
29. Karl Marx, *Capital*, volume 3 (London: Lawrence & Wishart, 1970), p. 820.
30. Michael Hardt and Antonio Negri, *Empire* (Cambridge, MA and London: Harvard University Press, 2000), p. 275. All further references will be given in the text.
31. Perhaps the most influential explanation of this shift in Marx's thought comes from Agnes Heller who argues that the philosopher moves from 'a paradigm of work' – in which labour is the defining mode of human species-existence – towards 'a paradigm of production' in which labour is devolved to technology. For Heller, Marx shifts from a view of human emancipation *through* labour to emancipation *from* labour via the expansion of industrialised production. In my view, however, Marx remains obscure on the question of what form human species-existence would take outside of the labour process: what *exactly* has been emancipated by production? See Agnes Heller, 'Paradigm of Production: Paradigm of Work', *Dialectical Anthropology* 6 (1981), pp. 71–9.
32. Richard Beardsworth, *Derrida and the Political* (London: Routledge, 1996), p. 161n. In Beardsworth's view, the dialectical logic of Marx's philosophy transforms the aporetic relation between the human and the technical into a '*politics* of human redemption' in which 'the history of matter is organised all too violently around the human', p. 161n. All further references will be given in the text.

3 Psyche

1. Sigmund Freud, *The Origins of Psychoanalysis: Letters to Wilhelm Fleiss, Drafts and Notes, 1887–1902* ed. by M. Bonaparte, A. Freud and E. Kris (New York: Basic Books, 1954), p. 173. All further references will be given in the text.
2. Jacques Derrida, 'Freud and the Scene of Writing', in *Writing and Difference* trans. by Alan Bass (London: Routledge and Kegan Paul, 1978), pp. 196–231. All further references will be given in the text. In the 40 years since it was written, Derrida's essay has been the subject of much exposition, commentary and debate not least by the philosopher himself. See, in particular, *Taking Chances: Derrida, Psychoanalysis and Literature* ed. by Joseph

H. Smith and William Kerrigan (Baltimore, MD: Johns Hopkins University Press, 1984); John Forrester, *The Seductions of Psychoanalysis: Freud, Lacan and Derrida* (Cambridge: Cambridge University Press, 1990); Christopher Johnson, *System and Writing in the Philosophy of Jacques Derrida* (Cambridge: Cambridge University Press, 1993), pp. 65–108; Jacques Derrida, *Archive Fever: A Freudian Impression* trans. by Eric Prenowitz (Chicago, IL: University of Chicago Press, 1996); and Maud Ellman, 'Deconstruction and Psychoanalysis', in *Deconstructions: A User's Guide* ed. by Nicholas Royle (London: Palgrave Macmillan, 2000), pp. 211–37.

3. Sigmund Freud, 'A Note upon the "Mystic Writing Pad"', in *The Complete Psychological Works*, volume 19 trans. by James Strachey (London: Hogarth Press, 1953–74), p. 231. All further references will be given in the text.
4. Friedrich Kittler, 'The World of the Symbolic – A World of the Machine', in *Literature, Media, Information Systems* ed. and intr. by John Johnston (Amsterdam: OPA, 1997), pp. 130–46, 134. All further references will be given in the text. In many ways, Kittler's own media histories can be seen as an attempt to articulate the technical conditions of psychoanalysis from the gramophone to cybernetics: 'A contemporary theory of consciousness which does not situate consciousness in the technical realm, as does Lacan, but has consciousness simultaneously transmit, store and calculate like a true behemoth without specifying the media or the technologies involved is merely a euphemism' (p. 133).
5. As Derrida observes some 30 years later in *Archive Fever*, the counter-factual question of how far the rise of psychoanalysis may have been influenced by the dominance of certain communication technologies – print, the postal service that delivered Freud's letters to Fleiss, not to mention the Mystic Writing Pad – and how that discipline may have developed if Freud and his followers had access to modern communication media such as fax, computers or email is an unanswerably vast one. In Derrida's words, 'Is the psychic apparatus *better represented* or is it *affected differently* by all the technical mechanisms for archivisation and for reproduction, for prostheses of so-called live memory, for simulacrums of living things which already are, and will increasingly be, more refined, complicated, powerful than the "mystic pad" (microcomputing, electronisation, computerisation, etc.)?' (p. 15).
6. Sigmund Freud, 'The Uncanny', in *The Complete Psychological Works*, volume 17 trans. by James Strachey (London: Hogarth Press, 1953–74), pp. 217–52. All further references will be given in the text.
7. Ernst Jentsch, 'On the Psychology of the Uncanny' trans. by Roy Sellars, *Angelaki* 2: 1 (1995), pp. 7–16.
8. E. T. A. Hoffman, *Tales of E. T. A Hoffman* trans. by. Leonard J. Kent and Elizabeth C. Knight (Chicago, IL: University of Chicago Press, 1969), p. 117.
9. To be sure, Freud's essay and, in particular, the relation between the uncanny and automata has been the subject of exhaustive commentary. See Hélène Cixous, 'Fiction and its Phantoms: A Reading of Freud's *Das Unheimliche* ("The Uncanny")', *New Literary History* 7 (1976), pp. 525–48; Neil Herz, 'Freud and the Sandman', in *The End of the Line: Essays on Psychoanalysis and the Sublime* (New York: Columbia University Press, 1985), pp. 97–121; Sarah Kofman, *Freud and Fiction* trans. by Sarah Wykes (Boston, MA: Northwestern University Press, 1991); Mladen Dolar, '*La femme-machine*', *New Formations*

23 (1994), pp. 43–54; David Wills, *Prosthesis* (Stanford, CA: Stanford University Press, 1995), pp. 92–129; and Nicholas Royle, *The Uncanny* (Manchester: Manchester University Press, 2003). In what follows, I am indebted to all of the above, whatever differences of interpretation arise.

10. See David Wills' remarkable *Prosthesis* for a discussion of the relationship between castration and automation. In Wills' account, the castration complex is not simply a fear of amputation but a fear of prosthetic replacement: 'If one projects the idea of substitution that is necessary for Freud to maintain that the eyes symbolise the penis, and if one reads the structure of detachability as necessarily implying that of replacement, then the fear of castration becomes in fact the fear of an inanimate otherness performing a human function' (p. 114).
11. Sigmund Freud, *Beyond the Pleasure Principle* trans. by James Strachey (New York: W. W. Norton, 1961).
12. Yet Herz goes on to argue that 'the awareness of the process of repetition is inseparable from the awareness of something between repeated' because 'there can be no such thing as sheer repetition' (p. 102). For Derrida, of course, there is *exactly* such a thing as sheer or originary repetition: this is what founds the immanent relation between the psyche and writing.
13. If Freud tends to represent this relation to death in terms of a premonition of *future* demise – the appearance of my double or doppelganger is famously a 'harbinger of death' (p. 235) – the phenomenon of repetition compulsion locates a non-living, inanimate or inorganic dimension that is already present *within* the structure of the psyche itself: the living psyche is thus never simply 'alive' in the first place.
14. Eric L. Santner, *On Creaturely Life: Rilke/Benjamin/Sebald* (Chicago, IL: University of Chicago Press, 2006), p. 191.
15. Jacques-Alain Miller, 'An Introduction to Seminars I and II: Lacan's Orientation Prior to 1953', in *Reading Seminars I and II: Lacan's Return to Freud* ed. by Richard Feldstein, Bruce Fink and Maire Jannus (New York: SUNY, 1996), pp. 3–38. In this essay, Miller gives an excellent overview of Lacan's intellectual itinerary up to 1953–54.
16. Manfred Frank, *Was ist Neostructuralismus?* (Frankfurt am Main: Suhrkamp, 1983), p. 383. Quoted in Kittler, 'The World of the Symbolic – A World of the Machine', pp. 131–2. All further references will be given in the text.
17. René Descartes, *Treatise on Man* trans. by Thomas Steele Hall (Cambridge, MA: Harvard University Press, 1972). In this short text, Descartes compares a human being with a hypothetical machine that operates like a clock or a hydraulic fountain but, as we saw in Chapter 1, the machine metaphor actually goes all the way back to Aristotle.
18. Jacques Lacan, *The Seminar of Jacques Lacan Book III: The Psychoses* trans. by Russell Grigg (London: Routledge, 1993), p.167.
19. Sigmund Freud, 'The Ego and the Id', in *The Complete Psychological Works*, volume 19 trans. by James Strachey (London: Hogarth Press). For Freud, the ego is merely a 'projection of the surface' of the body that contains mental life in the same way that skin contains the major organs (p. 26). In many ways, Lacan takes this concept of the exteriority of the ego to its logical conclusion: why shouldn't *any* material surface serve equally well to project the unconscious?

20. For Friedrich Kittler, who contests Frank's critique of the materialist definition of consciousness, Lacan returns to humanity at the end of the test only 'in order to verify a different function which Frank's criticism overlooks: the storage of data'. In my view, however, Kittler's apology merely defers the necessary arrival of the witness to a later stage: why is the human being invoked at all to 'verify' the storage of data if not because it is needed to grasp the relation between the datum, the reflection and the thing reflected? See Kittler, 'The World of the Symbolic – A World of the Machine', pp. 131–2.
21. Perhaps one way of re-formulating the Lacan-Pontalis exchange might be to see it as the beginning of the dispute between what have subsequently become known as the first and second waves of cybernetic theory over the role of the *observer*. On the one hand, the first generation of cyberneticists tends to ignore the observer of information systems: Weiner and Shannon seem to presume that they are merely recording an objective set of events that happen quite independently of them as if they were not there. On the other hand, the second generation of cyberneticists argues that the observer must be included within the information system: Bateson and, later, Maturana and Valera argue that the scientist does not merely record, but participates in, and shapes, a set of events that are in no way independent of him/her. In Bateson's terms, Lacan's question of whether the camera is a phenomenon of consciousness would thus be fundamentally misconceived: what is conscious would be the human *plus* the camera *plus* the environment. See Gregory Bateson, *Steps to an Ecology of Mind* (St Albans: Granada, 1973), p. 458.
22. In Lacan's account, 'A machine can scarcely do otherwise than feed itself... Machines which reproduce themselves have yet to be built, and have yet even to be conceived of – the schema of their symbolic has not even been established' (*Seminar II*, p. 54).
23. Alan Turing, 'Computing Machinery and Intelligence', *Mind* LIX 236 (1950), pp. 430–60. In fact, such an exchange was historically possible even if it never actually took place: Turing's famous essay was published some three years before Lacan's 1953/54 seminars.

4 Being

1. Martin Heidegger, 'The Question Concerning Technology', in *The Question Concerning Technology and Other Essays* trans. by William Lovitt (New York: Harper & Row, 1977), pp. 3–35, 4. All further references will be given in the text.
2. Robert C. Scharff and Val Dusek ed., *The Philosophy of Technology: The Technological Condition*, Blackwell Philosophy Anthologies (Oxford: Blackwell, 2003), p. 247.
3. Richard Rojcewicz, *The Gods and Technology: A Reading of Heidegger* (Albany: SUNY, 2006), p. 9. All further references will be given in the text.
4. Martin Heidegger, *Basic Questions of Philosophy: Selected 'Problems' of 'Logic'* trans. by Richard Rojcewicz and André Schuwer (Bloomington, IN: Indiana University Press, 1994), pp. 154–5. Quoted in Rojcewicz, p. 9. All further references will be given in the text.

5. Jacques Derrida, *Of Spirit: Heidegger and the Question* trans. by Geoffrey Bennington and Rachel Bowlby (Chicago, IL: University of Chicago Press, 1989), p. 10. All further references will be given in the text.
6. Martin Heidegger, *Introduction to Metaphysics* trans. by Gregory Fried and Richard Polt (New Haven, CT and London: Yale University Press, 2000), p. 15. All further references will be given in the text. See also Susan Schoenbohm, 'Heidegger's Interpretation of *Phusis*', in Heidegger's *Introduction to Metaphysics* ed. by Richard Polt and Gregory Fried (New Haven, CT and London: Yale University Press, 2000), pp. 143–60.
7. Martin Heidegger, 'On the Essence and Concept of *φύσις* in Aristotle's *Physics* B, I', pp. 183–230, 185. Heidegger's emphases. All further references will be given in the text.
8. Martin Heidegger, 'The Origin of the Work of Art' trans. by Albert Hofstadter, in *Poetry Language Thought* (New York: Harpercollins, 1971), pp. 15–86, 57–8. For Bernard Stiegler, '*Technē* qua *poiēsis* is thus submitted to the final cause that *phusis*, working through the efficient cause, constitutes' (*The Fault of Epimetheus*, p. 10). For me, Stiegler's language here is still somewhat too unHeideggerian – Heidegger consistently makes clear that there is no 'efficient causality' in Aristotle and he does not assign such absolute causality to the final cause.
9. Martin Heidegger, 'Plato's Doctrine of Truth' trans. by Thomas Sheehan, in *Pathmarks* ed. by William McNeil (Cambridge: Cambridge University Press, 1998), pp. 155–82. See also the *Introduction to Metaphysics*, pp. 138–98. It is Heidegger's claim that Plato reconceptualises the original notion of *eidos* as simply the 'look' or 'appearance' a thing has so that it becomes the 'traditional' metaphysical sense of *eidos* as the supersensory idea or form of the thing (*Introduction to Metaphysics*, p. 107). According to the former, this gesture marks the beginning of the metaphysical opposition between a being and its appearance that is quite foreign to the originary Greek thought of the phenomenon: 'Being essentially unfolds *as* appearing' (p. 107). For Heidegger, what is problematic in the Platonic text is not that Being becomes characterised by its appearance – Being must always unfold into an appearance – but rather that the idea or look comes to *absolutely determine* Being. If every being must have an appearance, what concerns Heidegger is that this appearance becomes the exclusive means of interpreting it: '[t]he visage offered by a thing, and no longer the thing itself, becomes what is decisive' (p. 195). See Laurence Paul Hemming, *Heidegger's Atheism: The Refusal of a Theological Voice* (Indiana, IN: University of Notre Dame Press, 2002) for an excellent account of this argument.
10. Martin Heidegger, *The Basic Problems of Phenomenology* trans. by Albert Hofstadter (Bloomington, IN: Indiana University Press, 1982), pp. 118–19.
11. Martin Heidegger, *Contributions to Philosophy (From Enowning)* trans. by Parvis Emad and Kenneth Maly (Bloomington, IN: Indiana University Press, 1999), p. 92. See also Jean-François Courtine, 'Les traces et le passages du Dieu dans les *Beiträge zur Philosophie* de Martin Heidegger', *Archivio di filosofia* (1994), pp. 519–38.
12. Jacques Derrida and Bernard Stiegler, *Echographies of Television* trans. by Jennifer Bajorek (Cambridge: Polity, 2002), p. 133. All further references will be given in the text.

13. David Wills, *Dorsality: Thinking Back Through Technology and Politics* (Minneapolis, MN and London: University of Minnesota Press, 2008), p. 34. For Wills, Heidegger's later philosophy of technology – with its famous Fourfold of gods, mortals, earth and world – similarly risks falling into an 'unexamined naturality or *phusis*'. In my view, though, the identification of *phusis* with unexamined naturality here would itself require further examination.
14. According to Michael E. Zimmerman, 'The Greek founders of metaphysics defined the being of entities in a proto-technological way. For them, "to be" meant "to be produced" '. Yet, everything depends upon the meaning of 'to be produced' here: 'production' does not mean technological manufacture but *poiēsis*, bringing-forth or nurture. In this sense, Heidegger is describing the 'becoming-natural' of technology rather than the 'becoming-technological' of nature. Michael E. Zimmerman, *Heidegger's Confrontation with Modernity: Technology, Politics, Art* (Bloomington, IN: Indiana University Press, 1990), p. xv.
15. Martin Heidegger, *Being and Time* trans. by John Macquarrie and Edward Robinson (Oxford: Blackwell, 1962), §15. All further references will be given in the text.
16. Jacques Derrida, 'Le main de Heidegger (*Geschlecht* II)', in *Psyché: Inventions de l'autre* (Paris: Galilée, 1987), pp. 415–71, 426.
17. To the best of my knowledge, the only really sustained attempt to offer such an ontically thick account of the autonomous life of the tool can be found in Graham Harman's attempt to re-think Heidegger's account of the tool via Bruno Latour's actor network theory. See, for example, Graham Harman, *Tool-being: Heidegger and the Metaphysics of Objects* (New York: Open Court, 2002).
18. Martin Heidegger, 'Martin Heidegger in Conversation with Richard Wisser' trans. by Lisa Harries, in *Martin Heidegger and National Socialism* ed. by Günther Neske and Emil Kettering (New York: Paragon House, 1990), p. 84. Translation modified.
19. Avital Ronnell, *The Telephone Book: Technology – Schizophrenia – Electric Speech* (Lincoln, NE and London: University of Nebraska Press, 1989).
20. See Martin Heidegger, 'Letter on Humanism' trans. by Frank A. Capuzzi, in *Basic Writings* (London: Routledge, 1978), pp. 217–65, 225, 243 and the *Zollikon Seminars: Protocols–Conversations–Letters* ed. by Medard Boss and trans. by Franz Mayr and Richard Askay (Evanston, IL: Northwestern University Press, 2001) for Heidegger's critique of the metaphysical assumptions underlying Marxian historical materialism, Freudian psychoanalysis and scientific positivism.
21. For Derrida, it is precisely this undecidability between ontological structure of revealability (*Offenbarkeit*) and ontic historical revelation (*Offenbarung*) which plagues Heidegger's phenomenology of religion: 'which one, first of all, will have signalled towards the other?'. Jacques Derrida, *Specters of Marx* (New York: Routledge, 1994), p. 168. In the same way, I think we can find an ontico-ontological undecidability in the philosophy of technology.
22. If the historically accepted view is that the epoch of modern science in the seventeenth century prepares the ground for the development of modern technology from the eighteenth century onwards, Heidegger argues

that, ontologically speaking, things are the other way around: what the philosopher calls the 'essence' of modern technology – the peculiar nexus of challenging, ordering and gathering at work in its view of nature – is already presupposed by the physical sciences (p. 22). See also the essay 'Science and Reflection', in *The Question Concerning Technology and Other Essays*, pp. 155–82.

23. To criticise Heidegger on the grounds that he simply idealises or abstracts empirical technology is to foist upon him a kind of Platonic metaphysical idealism that is foreign to his thought of the essence as *Wesen*. See Andrew Feenberg, 'Technology and Meaning', in *Questioning Technology* (New York and London: Routledge, 1999), pp. 183–99.
24. Giorgio Agamben, *The Open* (Stanford, CA: Stanford University Press, 2005), pp. 49–80. In Agamben's argument, Heidegger's ontological opposition between *Dasein* and the animal figures largely, but he does not discuss the later essays on technology in any detail.
25. Jacques Derrida, 'The Ends of Man', *Margins – Of Philosophy* trans. by Alan Bass (Brighton: Harvester Wheatsheaf, 1982), pp. 109–136, 28. In his career-long reading of Heidegger, Derrida identifies a number of symptoms of the philosopher's complicity with metaphysical anthropology: the privileging of *Dasein*'s proximity to itself and Being; the unique significance he attaches to the hand and to handcraft; the growing essentialisation of the thinking of spirit and, massively, the philosopher's reduction of the animal and animality.
26. Quentin Meillassoux, *After Finitude: An Essay on the Necessity of Contingency* trans. by Ray Brassier (London: Continuum, 2008). In Meillassoux's account, as we will see in Chapter 7, the absolute epistemological finitude described by post-Kantian philosophy is actually a misrecognised ontological state of radical contingency. All further references will be given in the text.
27. Werner Heisenberg, 'Über quantentheoretische Umdeutung kinematischer und mechanischer Beziehungen' *Zeitschrift für Physik*, 33 (1927), pp. 879–93. In this paradigm-shifting paper, Heisenberg famously argues that it is not possible to simultaneously describe certain pairs of observable quantum variables – position and momentum, energy and time – with a degree of accuracy that exceeds the limit imposed by Planck's constant (h).
28. Werner Heisenberg, 'Das Naturbild der heutigen Physik', in *Die Künste im techischen Zeitalter* (Munich: Oldenbourg, 1954), pp. 62–3.
29. Perhaps one of the largest questions left unanswered by the Copenhagen school is the problem of quantum cosmology: what possible observer could have measured the coming-into-being of the universe itself? If the Indeterminacy Principle demands that the universe can only exist because an external observer has collapsed its wave function, then the presumed absence of such a God-like observer necessarily implies the non-existence of the universe as well. In the Many Worlds Interpretation, this problem does not arise because no act of observation is required to bring the universe into actuality: there are an infinite number of actual universes.
30. Hugh Everett III, 'The Theory of the Universal Wavefunction', in Bryce DeWitt and R. Neill Graham, eds *The Many-Worlds Interpretation of Quantum Mechanics*, Princeton Series in Physics (Princeton, NJ: Princeton University Press, 1973), pp. 3–140.

31. In many ways, Alain Badiou's mathematical ontology makes this point in the most powerful way: Badiou argues that it is mathematics itself – in the particular form of the Zermelo–Fraenkel axiomatisation of Cantor's set-theory – that accomplishes the de-totalisation of Being which philosophers like Heidegger argue must be carried out from a poietic vantage point outside mathematics. See Alain Badiou, *Being and Event* trans. by Oliver Feltham (London: Continuum, 2005).

5 The Other

1. Jacques Derrida, *Memoires: For Paul de Man* trans. by Cecile Lindsay, Jonathan Culler and Eduardo Cadava (New York: Columbia University Press, 1986), p. 108. All further references will be given in the text.
2. Jacques Derrida, 'The Rhetoric of Drugs', in *Points… Interviews, 1974–1994* ed. by Elizabeth Weber and trans. by Peggy Kamuf (Stanford, CA: Stanford University Press, 1995), pp. 244–5, Derrida's emphases. All further references will be given in the text. See also Christopher Johnson, 'Derrida and Technology', in *Derrida's Legacies: Literature and Philosophy* ed. by Robert Eaglestone and Simon Glendinning (London: Routledge, 2008), pp. 54–65 for a recent assessment of Derrida's legacy to thinking on technics.
3. Jacques Derrida, *Edmund Husserl's* Origin of Geometry: *An Introduction* trans. by John P. Leavey, Jr (Lincoln, NE: University of Nebraska Press, 1978).
4. See Laurent Milesi, 'Almost Nothing at the Beginning: The Technicity of the Trace in Deconstruction', in *Language Systems: After Prague Structuralism* ed. by Louis Armand and Pavel Černovsky (Prague: Charles University Press, 2007), pp. 22–41 for what is, in my view, the best critical exegesis of the place of technicity in relation to Derrida's thinking of the trace. In what follows, I am very grateful to Milesi's outstanding essay notwithstanding any differences of emphasis and interpretation.
5. Jacques Derrida, 'Différance' trans. by Alan Bass, in *Margins – Of Philosophy* (Chicago, IL: University of Chicago Press, 1982), p. 17. All further references will be given in the text.
6. To be sure, Derrida argues that any attempt to reduce the general problematic of *arche*-writing to the *empirical* history of what he calls 'vulgar' writing succumbs to an uncritical scientism or positivism (*Of Grammatology*, pp. 74–86). It is, nonetheless, possible to detect a residual positivism in the many technological readings of deconstruction that seek to install *technē* as the proper, if hidden, name for *arche*-writing, the trace or *différance*. In Friedrich Kittler's *Gramophone, Film, Typewriter* (Stanford, CA: Stanford University Press, 1999) trans. by Geoffrey Winthrop-Young and Michael Wutz, for instance, we are told that the originary trace did not exist before the invention of the gramophone: 'The trace preceding all writing, the trace of pure difference still open between reading and writing, is simply a gramophone needle' (p. 33).
7. Jacques Derrida, *Positions* trans. by Alan Bass (Chicago, IL: University of Chicago Press, 1981), p. 64. In this series of interviews, Derrida is constantly (and somewhat aggressively) questioned about the relation between

deconstruction and dialectical materialism, but the philosopher is clearly reluctant to accept the comparison.

8. Jacques Derrida, 'Typewriter Ribbon: Limited Ink (2)', in *Material Events: Paul de Man and the Afterlife of Theory* ed. by Tom Cohen, Barbara Cohen, J. Hillis Miller and Andrzej Warminski (Minneapolis, MN: University of Minnesota Press, 2001), pp. 277–360, 353. To be fair, Derrida is actually referring to the theory of matter in the later Paul de Man's *Aesthetic Ideology*, but – notwithstanding the larger differences between the two – the description fits his own work as well. In many ways, though, Derrida's definition of 'matter' also risks turning deconstruction into little more than a new kind of nominalism. To what extent can 'matter' or reduced to a mere generic name or placeholder?
9. John Protevi, *Political Physics* (London: Continuum, 2001), p. 6. In Protevi's somewhat triumphalist version of recent philosophical history, Deleuze is nothing less than 'our Kant', an analogy that would seem (though Protevi does not say this) to relegate Derrida to the position of a pre-critical predecessor – Leibniz perhaps?
10. Jacques Derrida and Bernard Stiegler, *Echographies of Television* trans. by Jennifer Bajorek (Cambridge: Polity Press, 2002), p. 133. All further references will be given in the text.
11. Jacques Derrida, 'The Principle of Reason: The University in the Eyes of its Pupils' trans. by Catherine Porter and Edward P. Morris, *Diacritics* (Fall, 1983), pp. 6–20, 14.
12. See Joanna Hodge, *Derrida on Time* (London: Routledge, 2007) for a remarkable conceptual genealogy of Derrida's philosophy of time from his earliest engagements with Husserl to his later engagements with de Man and Stiegler.
13. Edmund Husserl, *The Phenomenology of Internal Time Consciousness* trans. by J. Churchill (Bloomington, IN: Indiana University Press, 1964), pp. 50–1.
14. Jacques Derrida, *Speech and Phenomena and Other Essays on Husserl's Theory of Signs* ed. and trans. by David Allison (Evanston, IL: Northwestern University Press, 1973).
15. Mark B. N. Hansen, 'Realtime Synthesis and the Différance of the Body: Technocultural Studies in the Wake of Deconstruction', *Cultural Machine* 6 (2004) (http://www.culturemachine.net/index. php/cm/article/view/9/8) (accessed 13 November 2009).
16. Richard Beardsworth, 'Towards a Critical Culture of the Image: J. Derrida and B. Stiegler's *Echographies de la television*', *Tekhnema: Journal of Philosophy and Technology* 4 (1998) (http:// tekhnema.free.fr/4Beardsworth.html) (accessed 12 November 2009). All further references will be given in the text.
17. Jacques Derrida, 'Some Statements and Truisms about Neo-logisms, Newisms, Postisms, Parasitisms and Other Small Seisisms' trans. by A. Tomiche, in *The States of 'Theory': History, Art and Critical Discourses* ed. by David Carroll (New York: Columbia University Press, 1991), p. 85.
18. Geoffrey Bennington, 'Emergencies', *Oxford Literary Review* 18 (1996), pp. 175–216. In this typically incisive review essay of the work of Bernard Stiegler and Richard Beardsworth, Bennington accuses them of reducing a quasi-transcendental claim about originary technicity into a positivist history of empirical technics. See Section 4 of this chapter for an evaluation of this argument. All further references will be given in the text.

19. Jacques Derrida, *Paper Machine* trans. by Rachel Bowlby (Stanford, CA: Stanford University Press, 2005).
20. Jean-Luc Nancy, *La Déclosion: Déconstruction du christianisme 1* (Paris: Editions de Galilée, 2005). All further references will be given in the text. Translations mine.
21. Jacques Derrida, *On Touching – Jean-Luc Nancy* trans. by Christine Irizarry (Stanford, CA: Stanford University Press. 2005), pp. 59–60, 220 All further references will be given in the text. To be sure, the number of occasions where Derrida draws a parallel between Heidegger's *destruktion* and Luther's *destructio* are almost too numerous to mention. In many ways, Derrida's line of argument is merely the latest example of a theological contextualisation of Heidegger that dates back to the work of Rudolf Bultmann. See *The New Testament and Mythology and Other Basic Writings* (Philadelphia, PA: Fortress, 1984).
22. Jean-Louis Chrétien, *L'appel et la réponse* (Paris: Minuit, 1992).
23. Dominique Janicaud, Jean-François Courtine, Jean-Louis Chrétien, Michel Henry, Jean-Luc Marion and Paul Ricoeur, *Le tournant théologique de la phénoménologie française* (Paris: L'Eclat, 1991).
24. André Leroi-Gouhran, *Gesture and Speech* (Cambridge, MA: MIT Press), p. 18. In fact, Derrida's *De la grammatologie* began life as a review article, published in *Critique* in 1965 and 1966, of three books including Leroi-Gouhran's own *Le geste et le parole*. See Arthur Bradley, *Derrida's* Of Grammatology: *A Philosophical Guide* (Bloomington, IN: Indiana University Press, 2008) for an account of the gestation of Derrida's book.
25. Jacques Derrida, 'Faith and Knowledge: The Two Sources of "Religion" at the Limits of Reason Alone' trans. by Samuel Weber, in *Religion* ed. by Jacques Derrida and Gianni Vattimo (Stanford, CA: Stanford University Press, 1998), pp. 1–78, 18. All further references will be abbreviated in the text.
26. Jacques Derrida, *The Gift of Death* trans. by David Wills (Chicago, IL: University of Chicago Press, 1995), p. 49. See Arthur Bradley, *Negative Theology and Modern French Philosophy* (London: Routledge, 2004) for an account of the return of the religious in Derrida, Michel de Certeau, Jean-Luc Marion, Michel Foucault and Julia Kristeva.
27. Kevin Hart, *The Trespass of the Sign: Mysticism and Deconstruction* (Cambridge: Cambridge University Press, 1989); John D. Caputo, *The Prayers and Tears of Jacques Derrida: Religion Without Religion* (Bloomington, IN: Indiana University Press, 1997); Hent de Vries, *Philosophy and the Turn to Religion* (Baltimore, MD: Johns Hopkins University Press, 1999). For de Vries, 'God' is quite simply the 'best' name for the self-erasing trace or aporia of origin with which deconstruction is famously concerned (pp. 94, 355).
28. Bernard Stiegler, 'Derrida and Technology: Fidelity at the Limits of Deconstruction and the Prosthesis of Faith' trans. by Richard Beardsworth, in *Jacques Derrida and the Future of the Humanities* ed. by Tom Conley (Cambridge: Cambridge University Press, 2002), pp. 238–70; Martin Hägglund, *Radical Atheism: Derrida and the Time of Life* (Stanford, CA: Stanford University Press, 2008); Richard Beardsworth, 'The Messianic Now: A Secular Response', in *The Politics to Come: Power, Modernity and the Messianic* ed. by Arthur Bradley and Paul Fletcher (London: Continuum, 2010), pp. 13–25.

29. In many ways, all talk of Marxian or Heideggerian 'turns' in Derrida's work – whether ethical, political or theological – risks creating hostages to fortune and any cursory reader of his work will be able to pick holes in the kind of historical taxonomy set out above. As numerous critics have pointed out, Derrida's supposed 'turn' to a quasi-Levinasian ethics and politics in the later work is actually present at the very beginning, when, for example, the relation to the trace is explicitly tied to the relation to the other in the *Grammatology* (p. 46). Perhaps Derrida's so-called theological turn was also pre-empted at the very start by Brice Parain's incorrect, but with hindsight, very revealing, allegation that deconstruction was a negative theology in response to the very first oral presentation of the essay 'La Différance' in 1968: 'It [*différance*] is the source of everything and one cannot know it. It is the God of negative theology.' See Jacques Derrida, 'The Original Discussion of Différance', in *Derrida and Différance* ed. by Robert Bernasconi and David Wood (Evanston, IL: Northwestern University Press, 1988), p. 84.
30. Christopher Johnson, 'I-You-We, Robot', in *Technicity* ed. by Arthur Bradley and Louis Armand (Prague: Charles University Press, 2006), pp. 101–28.

6 Time

1. Bernard Stiegler, *Technics and Time 1: The Fault of Epimetheus* (Stanford, CA: Stanford University Press, [1994] (1998); *La Technique et le temps 2: La Désorientation* (Paris: Galilée, 1996); and *La Technique et le temps 3: Le Temps du cinema et la question du mal-être* (Paris: Galilée, 2001). In the *Avertissement* to the third volume, Stiegler mentions a fourth and possibly even a fifth volume which, at the time of writing, have yet to appear: *Symboles et diaboles ou les guerres des spirites* (which treats of Plato and Derrida) and *Le défaut qu'il faut*. All further references to these works will be given in the text.
2. It is typical, however, that Stiegler also identifies a technical blindspot even within Simondon's theory of individuation: what Simondon does not recognise is the extent to which the pre-individual milieu out of which every individual identity is formed is *itself* retained through epiphylogenetic systems. See Bernard Stiegler, *De la misère symbolique 1: L'époque hyperindustrielle* (Paris: Galilée, 2004), p. 106.
3. Geoffrey Bennington, 'Emergencies', *Oxford Literary Review* 18 (1996), pp. 175–216. In Chapter 5, however, we saw that Stiegler makes very clear that he is not reducing what Derrida calls *différance* to some kind of material or technological *positum*, but doing something more subtle: we would have no *access* to the logic of the supplement of which Derrida speaks except *through* the capacity for temporalisation embodied in empirical technics. See Bernard Stiegler, 'Derrida and Technology: Fidelity at the Limits of Deconstruction and the Prosthesis of Faith' trans. by Richard Beardsworth, in *Jacques Derrida and the Future of the Humanities* ed. by Tom Conley (Cambridge: Cambridge University Press, 2002), pp. 238–70.
4. Mark B. N. Hansen, 'The Time of Affect, or Bearing Witness to Life', *Critical Inquiry*, 30: 3 (2004), pp. 584–626. In many ways, however, this is a slightly one-sided reading of Stiegler's work because the latter – as Hansen recognises in later essays – is always insistent on the organic embodiment of

consciousness: everything begins with the feet and the hands. See Bernard Stiegler, *Des Pieds et des mains* (Paris: Bayard Culture, 2006).

5. Richard Beardsworth, 'From a Genealogy of Matter to a Politics of Memory: Stiegler's Thinking of Technics', *Tekhnema: Journal of Philosophy and Technology* 2 (1995) (http://tekhnema.free.fr/contents4.html) (accessed 1 October 2009). In Beardsworth's account, Stiegler's philosophy repeats Derrida's deconstruction of the opposition between the transcendental and the empirical but also seeks to transform that gesture by supplying a history of what precedes oppositional logic.
6. Bernard Stiegler, *Philosopher par accident: Entretiens avec Elie During* (Paris: Galilée, 2004), p. 42. In Stiegler's words, 'mortals are beings without qualities, which is the difference between them and animals, for whom the distribution of qualities is carried out smoothly' (my translation). All further references will be given in the text.
7. Bernard Stiegler, 'Leroi-Gouhran: L'Inorganique organisé', *Cahiers de Médiologie*, 6, 2e Semestre (1998), p. 191 (my translation).
8. Richard Beardsworth, 'Thinking Technicity', *Cultural Values*, 2: 1 (1998), pp. 70–86. In this sense, Stiegler's view of originary technicity is significantly narrower than, say, that of Gilles Deleuze or Jean-François Lyotard: Lyotard insists that DNA is already a form of technique inasmuch as it is capable of accumulating and adapting the experience of previous life forms. See *The Inhuman: Reflections on Time* trans. by Geoffrey Bennington and Rachel Bowlby (Cambridge: Polity, 1991), p. 52.
9. Mark B. N. Hansen, 'Media Theory', *Theory, Culture & Society* 23: 2–3 (2006), pp. 297–306, 301 (Hansen's emphasis). All further references will be given in the text.
10. Jacques Derrida, Sous la direction de Marie-Louise Mallet, *L'animal autobiographique: Autour de Jacques Derrida* (Paris: Galilée, 1999); Cary Wolfe ed. *Zoontologies: The Question of the Animal* (Minneapolis, MN: Minnesota University Press, 2003); Giorgio Agamben, *The Open: Man and Animal* trans. by Kevin Attell (Stanford, CA: Stanford University Press, 2005). All further references will be given in the text.
11. To be sure, animal tool-use and fabrication are generally acknowledged to be widespread in natural history. It is now well documented that wild chimpanzees in Africa use a 'kit' of simple and complex tools for foraging, socialising and cleaning themselves. For chimpanzees, tool-use performs the same temporalising function that it did for early man: a series of different tools are fashioned and deployed in the correct order in order to achieve a single goal like extracting honey from a beehive. Finally, it has also been argued such tools constitute examples of technical or 'epiphylogenetic' memory: this kind of experience is undoubtedly too complex to be learnt from scratch by every chimpanzee and is acquired through a material culture. See W. C. McGrew, *Chimpanzee Material Culture: Implications for Human Evolution* (Cambridge: Cambridge University Press, 1992).
12. Perhaps one might recall Michel Foucault's Collège de France lectures here which date the birth of the human species to the epoch of liberal bio-politics: men ceased to be called 'mankind (*le genre humaine*)' and began to be called 'the human species (*l'espèce humaine*) or biological being (*être biologique*)'. For Foucault, in other words, humanity becomes a purely biological being

defined in terms of its species properties. In my own view, one of the comparative blindspots in *Technics and Time* is Stiegler's failure to address the bio-political as well as purely philosophical dimensions at work in the *naturalisation* of human existence as species-being and the bio-political possibilities afforded by its technical transformation back into 'humankind'. See Michel Foucault, *Security, Territory, Population: Lectures at the Collège de France 1977–78* trans. Graham Burchell (Basingstoke and New York: Palgrave Macmillan, 2007).

13. If Stiegler is now regularly convicted of a residual anthropologism, it is only fair to note that the philosopher himself totally refutes the charge: what he calls the technical 'pro-gramme' applies equally to all organic beings, from flowers to human beings (Bernard Stiegler, personal communication with the author). In Stiegler's more recent work, I would admit that it is certainly possible to detect a shift away from the neo-Heideggerian emphases of the first volume of *Technics and Time.*
14. Jean-Michel Salanskis, 'Ecce Faber', *Les Temps Modernes* 55: 608 (2000), pp. 235–78.
15. One revealing symptom of this valorisation is what we will see to be his consistent stress on *temporisation*, on delaying or slowing down the speed of the technical synthesis of Real Time, until it coincides with the perceptual synthesis that makes criticism possible.
16. For Stiegler, '[w]hat we call "real time" is thus not time; it is perhaps even the de-temporalisation of time, or the occultation of time; and yet it is, nonetheless, time, "gained" industrially, which is also to say lost', *Désorientation*, p. 62. (my translation).
17. Roland Barthes, *Camera Lucida: Reflections on Photography* trans. by R. Howard (New York: Hill and Wang, 1981). In Barthes' account, the '*this was* [*le ça a été*]' – of an analogue photograph is an essential component of what he famously calls its *punctum.*
18. In Salanskis' review essay, he accuses Stiegler of conflating two quite different Real Time technologies with two quite different purposes. On the one hand, there are 'Real Time' digital computing systems (extensively used in the financial industry) that operate quite independently of, and far more quickly than, human time consciousness. On the other, there are 'Real Time' media (such as live television programmes) that, whilst operating at a speed beyond the threshold of perception, are still explicitly designed to produce a phenomenon – the 'live event' – which is accessible to perceptual consciousness. See 'Ecce Faber', *Les Temps Modernes* 55: 608 (2000), p. 274.
19. Bernard Stiegler, *Prendre soin: De la jeunesse et des générations* (Paris: Flammarion, 2008). But who and/or what decides the duration of the 'short-term' and the 'long-term' in this account?
20. Jean François Lyotard, '*Logos* and *Techne*, or Telegraphy', in *The Inhuman: Reflections on Time* (pp. 47–57, 51). In Lyotard's words, 'Stiegler is right to insist on the need to make [contemporary techno-culture's] specific mode of inscription (and therefore of memorisation) available to individuals. School used to teach future citizens how to write. What institution has responsibility for teaching telegraphy? Can the ideal produced by such a citizen still be the citizen? Is an institution for the telegraphisation of humans even possible? Is the idea of an institution not linked to the State and to writing and reading? And thus to the ideal of a political body? It is abundantly clear in

any case that States are not the agencies in control of the general process of the new telegraphic bleaching, which in principle goes well beyond them.'

7 Death

1. James Joyce, *Portrait of the Artist as a Young Man* (London: Penguin, 2000), p. 186. In May 2010, the genetic biologist Craig Venter announced that his research team had created what he called the first 'synthetic cell': a microbe that thrives and replicates – lives – under the control of a genome painstakingly constructed in a laboratory. This single-celled organism had a number of 'watermarks' written into its DNA to identify it as synthetic and help trace its descendants back to their creator, including the above quotation from Joyce. See Ian Sample, 'Craig Ventner Creates Synthetic Life Form', *Guardian*, 20 May 2010.
2. See Marvin Minsky, *The Society of Mind* (London: Simon and Schuster, 1987); Hans Moravec, *Robot: Mere Machine to Transcendent Mind* (Oxford: Oxford University Press, 1999); Nick Bostrom, 'A History of Transhuman Thought', *Journal of Evolution and Technology* 14: 1 (2005), pp. 1–25; Ray Kurzweil, *The Singularity is Near* (London: Duckworth, 2006).
3. N. Katherine Hayles, *How We Became Posthuman: Virtual Bodies in Cybernetics, Literature and Informatics* (Chicago, IL and London: University of Chicago Press, 1999), p. 1. All further references will be given in the text.
4. See, for example, Donna Haraway, 'A Cyborg Manifesto: Science, Technology and Socialist-Feminism in the Late Twentieth Century', in *Simians, Cyborgs and Women: The Reinvention of Nature* (New York: Routledge, 1991), pp. 149–81; Neil Badmington ed., *Posthumanism: A Reader* (London: Palgrave, 2000); Elaine L. Graham, *Representations of the Post/Human: Monsters, Aliens and Others in Popular Culture* (Manchester: Manchester University Press, 2002); Bruce Clark, *Posthuman Metamorphosis: Narrative and Systems* (New York: Fordham University Press, 2008); Cary Wolfe, *What is Posthumanism?* (Minneapolis, MN: University of Minnesota Press, 2010). All further references will be given in the text.
5. Eugene Thacker, *Biomedia* (Minneapolis, MN: University of Minnesota Press, 2004); Nikolas Rose, *The Politics of Life Itself: Biomedicine, Power, and Subjectivity in the Twenty-First Century* (Princeton, NJ: Princeton University Press, 2006); Melinda Cooper, *Life as Surplus: Biotechnology and Capitalism in the Neoliberal Era* (Washington, DC: University of Washington Press, 2008).
6. Catherine Malabou, *What Should We Do With Our Brain?* trans. by Sebastian Rand (New York: Fordham University Press, 2008); Emmaunel Levinas, *Totality and Infinity: An Essay on Exteriority* trans. by Alphonso Lingis (Pittsburgh, PA: Duquesne University Press, 1969); Gilles Deleuze, *Francis Bacon: The Logic of Sensation* trans. and intr. by Daniel W. Smith (Minneapolis, MN: University of Minnesota Press, 2003); Jacques Derrida, *On Touching – Jean-Luc Nancy* trans. by Christine Irizarry (Stanford, CA: Stanford University Press, 2002); Bernard Stiegler, *Des Pieds et des mains* (Paris: Bayard Culture, 2006); Daniel Heller-Roazen, *The Inner Touch: Archaeology of a Sensation* (Cambridge, MA: MIT Press and Zone Books, 2007); Maurice Merleau-Ponty, *Phenomenology of Perception* trans. by Colin Smith (New York: Humanities Press, 1962);

Didier Anzieu, *The Skin Ego* trans. by Chris Turner (New Haven, CT: Yale University Press, 1989); Julia Kristeva, *Revolution in Poetic Language* trans. by Margaret Waller (New York: Columbia University Press, 1984); Luce Irigaray, *Speculum of the Other Woman* trans. by Gillian C. Gill (New York: Columbia University Press, 1985); David Wills, *Dorsality: Thinking Back Through Technology and Politics* (Minneapolis, MN and London: University of Minnesota Press, 2008); Jacques Lacan, *The Four Fundamental Concepts of Psychoanalysis* trans. by Alan Sheridan (Harmondsworth: Penguin, 1977); Slavoj Žižek, *The Indivisible Remainder* (London: Verso, 1996).

7. To be sure, the turn to the body has produced a vast literature but see the Introduction to Phillip Thurtle and Robert Mitchell eds, *Data Made Flesh: Embodying Information* (New York: Routledge, 2004) for a good overview of the field.
8. Timothy Clark, *The Poetics of Singularity: The Counter-Culturalist Turn in Heidegger, Derrida, Blanchot and the Later Gadamer* (Edinburgh: Edinburgh University Press, 2005), p. 26. In Clark's polemical but important account, contemporary literary studies are convicted of presupposing a liberal identity politics which is institutionally American.
9. Slavoj Žižek, *On Belief* (London: Routledge, 2001), pp. 153–4, n. 27. For Žižek, Hayles is guilty of constructing a 'crude opposition' between the embodied posthuman subject and the liberal autonomous subject that underestimates the extent to which the latter 'already IS "post-human"': Enlightenment philosophy not only begins to think of the human as a mechanism (Julien le Mettrie's *l'homme machine*) but also subtracts the subject from any finite embodiment or personhood (the Kantian subject of transcendental apperception).
10. According to Hayles, what distinguishes humanism from posthumanism is that the former valorises a Hobbesian and Lockean 'state of nature' in which the individual originally resides whereas the latter dispenses with the idea of a 'natural' self and replaces it with a 'material-informational entity whose boundaries undergo continuous construction and reconstruction' (*How We Became Posthuman*, p. 3). In my own view, though, such a static opposition underestimates the dynamic and radical transformations in 'the nature of nature' that attended liberal modernity – transforming both natural law and human nature from a providentially endowed order into an endlessly malleable and negotiable system whose only norm is the promotion of human welfare – and which prepared the ground for exactly the kind of posthumanist model of subjectivity Hayles wishes to valorise. See, for example, Ian Hunter, *Rival Enlightenments: Civil and Metaphysical Philosophy in Early Modern Germany* (Cambridge: Cambridge University Press, 2001).
11. Charles Taylor, *A Secular Age* (Cambridge, MA and London: Harvard University Press, 2007). In Taylor's account, the 'buffered' self of post-Lockean liberalism needs to be supplemented by a 'porous' communitarian self which is constituted by, and permanently open to, the possibility of self-transcendence. For Taylor, this transcendence finds its paradigm in religion rather than technology.
12. Mark B. N. Hansen, *Embodying* Technesis*: Technology Beyond Writing*; *New Philosophy for New Media* (Cambridge, MA: MIT Press, 2004); *Bodies in*

Code: Interfaces with Digital Media (New York: Routledge, 2006). All further references will be given in the text.

13. If Hansen remains firm in his belief that – for all its cognitive sophistication – technology is not autonomous of the human, the theorist does occasionally concede that this conviction is mortgaged to a historically finite prosthetic idea of technology that could all too conceivably be rendered bankrupt in the near future. In Hansen's words, 'if "we" do one day succeed in artifactually generating artificial life, a truly new evolutionary lineage, or rather, if out of the process of cosmic life in which we participate, such an artifactual genesis of artificial life emerges, it will be one that is no longer correlated – via the species-constituting transduction of the human and technics – with the human, and thus one that may well lay claim to true autonomy'. See Mark B. N. Hansen, 'Media Theory', *Theory, Culture & Society* 23: 2–3 (2006), pp. 297–306, 302.
14. Perhaps another sign of the somewhat polemical nature of Hansen's critique is that he neglects to mention that Lyotard can quite legitimately be seen as a thinker of embodiment from *Libidinal Economy* all the way to the *Inhuman*. In 'Can Thought Go On Without a Body?', for example, Lyotard attacks the Cartesian dualism of mind and body by insisting upon the necessarily affective and gendered basis of cognition: 'Thought is inseparable from the phenomenological body', p. 23.
15. Henri Bergson, *Matter and Memory* trans. by N. M. Paul and W. S. Palmer (New York: Zone Books, 1988), pp. 17–19.
16. Stephen Dougherty, 'New Media's Body', in *Language Systems: After Prague Structuralism* ed. by Louis Armand (Prague: Literaria Pragensia), pp. 65–78, 71. All further references will be given in the text. In my reading of Hansen, I am deeply indebted to Dougherty's excellent essay.
17. Jacques Derrida, *On Touching – For Jean-Luc Nancy*, pp. 183–215. For Derrida, Merleau-Ponty's body-schema is – far from being the embodied antidote to the transcendental attitude that Hansen claims – as 'unifying and synthetic as a Kantian schema' (p. 214).
18. See James Williams, *Lyotard and the Political* (London: Routledge, 2000), pp. 119–34. All further references will be given in the text. In the paragraphs that follow, I am indebted to Williams' incisive discussion of the later Lyotard.
19. Immanuel Kant, 'Analytic of the Sublime', in *Critique of Judgement* trans. by Werner S. Pluhar (Indianapolis, IN and Cambridge: Hackett, 1987), pp. 97–201.
20. Yu Liu, 'The Beautiful and the Sublime: Kant's Paradise Lost and Paradise Regained', *Studies in Romanticism* 42: 3 (2003), p. 194.
21. Jean-François Lyotard, 'Kant Notice 4', in *The Differend: Phrases in Dispute* trans. by Georges Van Der Abbeele (Minneapolis, MN: University of Minnesota Press, 1988), pp. 161–9. See also *Lessons on the Analytic of the Sublime* trans. by Elizabeth Rottenberg (Stanford, CA: Stanford University Press, 1994).
22. Immanuel Kant, 'A Renewed Attempt to Answer the Question "Is the Human Race Continually Improving?" ', in *Political Writings* ed. and intro. by Hans Reiss and trans. by H. B. Nisbet (Cambridge: Cambridge University Press, 2nd edition, 1991), pp. 176–90.

23. For Burke in the *Philosophical Enquiry into the Origin of Our Ideas of the Sublime and the Beautiful*, the true measure of the sublime is the feeling – a certain degree of horror without the sense of actual physical danger – it invokes, but Kant would reject this psychological judgement on the grounds that it denies the aesthetic judgement any universal jurisdiction. It merely states what *is contingently experienced as* sublime, in other words, rather than what *necessarily or ought to be adjudged* sublime. In fact, Kant might level the same objection to Lyotard that he makes to Burke: it is necessary to lift aesthetic judgements 'out [of the reach of] empirical psychology, in which they would otherwise remain buried amongst the feelings of gratification and pain (accompanied only by the epithet of being a *more refined* feeling)', *Critique of Judgement*, p. 125.
24. Manuel De Landa, *War In the Age of Intelligent Machines* (New York: Zone, 1991), pp. 1–10. All further references will be given in the text. In classical rhetoric, *prosopopoeia* is a device in which a speaker or writer communicates to the audience by speaking as another person or object.
25. Manuel de Landa, *A Thousand Years of Non-Linear History* (New York: Zone, 1997); *Intensive Science and Virtual Philosophy* (London: Continuum, 2005); and *A New Philosophy of Society: Assemblage Theory and Social Complexity* (London: Continuum, 2006). All further references will be given in the text. In many ways, the father of this school of history was Fernand Braudel in seminal works like *La Méditerranée et le Monde Méditerranéen a l'époque de Philippe II* (1949) (Paris: Librairie, Armand Colin, 2nd edition, 1967).
26. Jean-Luc Nancy, *A Finite Thinking* trans. by Simon Sparks (Stanford, CA: Stanford University Press, 2003), pp. 25–6.

Bibliography

Agamben, Giorgio, *Homo Sacer: Sovereign Power and Bare Life* trans. by Daniel Heller-Roazen (Stanford, CA: Stanford University Press, 1998).

———, *The Open: Man and Animal* trans. by Kevin Attell (Stanford, CA: Stanford University Press, 2005).

Althusser, Louis and Balibar, Etienne, *Reading Capital* trans. by Ben Brewster (New York: New Left Books, 1970).

Anzieu, Didier, *The Skin Ego* trans. by Chris Turner (New Haven, CT: Yale University Press, 1989).

Aristotle, *Politics* trans. by T. A. Sinclair rev. by Trevor J. Saunders (Harmondsworth: Penguin, 1981).

———, *Physics* trans. by Robin Waterfield (Oxford: World's Classics, 1999).

Armand, Louis, 'Technics and Humanism', in *Event States: Discourse, Time, Mediality* (Prague: Charles University Press, 2007), pp. 225–61.

Axelos, Kostas, *Marx, penseur de la technique: De l'aliénation de l'homme à la conquête du monde* (Paris: Editions de Minuit, 1967).

Badiou, Alain, *Being and Event* trans. by Oliver Feltham (London: Continuum, 2005).

Badmington, Neil ed., *Posthumanism: A Reader* (London: Palgrave Macmillan, 2000).

Bajorek, Jennifer, 'Animadversions: *Tehkne* after Capital/Life after Work', *Diacritics* 33: 1 (2003), pp. 42–59.

Balansard, Anne, *Techné dans les dialogues de Platon* (Sankt Augustin: Academia Verlag, 2001).

Balibar, Etienne, *La Philosophie de Marx* (Paris: La Découverte, 1993).

Barthes, Roland, *Camera Lucida: Reflections on Photography* trans. by R. Howard (New York: Hill and Wang, 1981).

Bateson, Gregory, *Steps to an Ecology of Mind* (St Albans: Granada, 1973).

Beardsworth, Richard, 'From a Genealogy of Matter to a Politics of Memory: Stiegler's Thinking of Tecnnics', *Tekhnema: Journal of Philosophy and Technology* 2 (1995) (http://tekhnema.free.fr/contents4.html), accessed 1 October 2009.

———, *Derrida and the Political* (London: Routledge, 1996).

———, 'Thinking Technicity', *Cultural Values* 2: 1 (1998), pp. 70–86.

———, 'Towards a Critical Culture of the Image: J. Derrida and B. Stiegler's *Echographies de la télévision*', *Tekhnema: Journal of Philosophy and Technology* 4 (1998) (http.tekhnema.free.fr/4Beardsworth.html), accessed 1 October 2009.

———, 'The Messianic Now: A Secular Response', in *The Politics to Come: Power, Modernity and the Messianic* ed. by Arthur Bradley and Paul Fletcher (London: Continuum, 2010), pp. 13–25.

Bennington, Geoffrey, 'Emergencies', *Oxford Literary Review* 18 (1996), pp. 175–216.

Bergson, Henri, *Matter and Memory* trans. by N. M. Paul and W. S. Palmer (New York: Zone Books, 1988).

Bimber, Bruce, 'Three Faces of Technological Determinism', in *Does Technology Drive History?* ed. by M. R. Smith and L. Marx (Cambridge, MA: MIT Press, 1994), pp. 79–100.

Blackmore, Susan, *The Meme Machine* (Oxford: Oxford University Press, 2000).

Bostrom, Nick, 'A History of Transhuman Thought', *Journal of Evolution and Technology* 14: 1 (2005), pp. 1–25.

Bradley, Arthur, *Negative Theology and Modern French Philosophy* (London: Routledge, 2004).

———, *Derrida's* of Grammatology: *A Philosophical Guide* (Bloomington, IN: Indiana University Press, 2008).

——— and Fletcher, Paul eds, *The Politics to Come: Power, Modernity and the Messianic* (London: Continuum, 2010).

Bultmann, Rudolf, *The New Testament and Mythology and Other Basic Writings* (Philadelphia, PA: Fortress, 1984).

Burke, Edmund, *A Philosophical Enquiry into the Origin of Our Ideas of the Sublime and Beautiful* (Oxford: World's Classics, 1998).

Canguilhem, Georges, *La Connaissance de la vie* (Paris: Vrin, 1985).

———, 'Machine and Organism', trans. by Mark Cohen and Randall Cherry, in *Incorporations* ed. by Jonathan Crary and Sanford Kwinter (New York: Zone Books, 1992), pp. 44–69.

Caputo, John D., *The Prayers and Tears of Jacques Derrida: Religion Without Religion* (Bloomington, IN: Indiana University Press, 1997).

Cheah, Pheng, 'Human Freedom and the Technic of Nature', *Differences: A Journal of Feminist Cultural Studies* 14: 2 (2003), pp. 1–26.

Chrétien, Jean-Louis, *L'appel et la réponse* (Paris: Minuit, 1992).

Cixous, Hélène, 'Fiction and Its Phantoms: A Reading of Freud's *Das Unheimliche* ("The Uncanny")', *New Literary History* 7 (1976), pp. 525–48.

Clark, Andy, *Natural-Born Cyborgs: Minds, Technologies and the Future of Human Intelligence* (New York: Oxford University Press, 2003).

Clark, Bruce, *Posthuman Metamorphosis: Narrative and Systems* (New York: Fordham University Press, 2008).

Clark, Timothy, 'Deconstruction and Technology', in *Deconstructions: A User's Guide* ed. by Nicholas Royle (London: Palgrave Macmillan, 2000), pp. 238–57.

———, *The Poetics of Singularity: The Counter-Culturalist Turn in Heidegger, Derrida, Blanchot and the Later Gadamer* (Edinburgh: Edinburgh University Press, 2005).

Cooper, Melinda, *Life as Surplus: Biotechnology and Capitalism in the Neoliberal Era* (Washington, DC: University of Washington Press, 2008).

Courtine, Jean-François, 'Les traces et le passages du Dieu dans les *Beiträge zur Philosophie* de Martin Heidegger', *Archivio di filosofia* (1–3) (1994), pp. 519–38.

Dawkins, Richard, *Climbing Mount Improbable* (London: Penguin, 1996).

De La Mettrie, Julien, *Man a Machine* trans. by G. C. Bussey (Chicago, IL: Open Court, 1912).

De Landa, Manuel, *War in the Age of Intelligent Machines* (New York: Zone, 1991).

———, *A Thousand Years of Non-Linear History* (New York: Zone, 1997).

———, *Intensive Science and Virtual Philosophy* (London: Continuum, 2005).

———, *A New Philosophy of Society: Assemblage Theory and Social Complexity* (London: Continuum, 2006).

Deleuze, Gilles, *Francis Bacon: The Logic of Sensation* trans. and intr. by Daniel W. Smith (Minneapolis, MN: University of Minnesota Press, 2003).

——— and Guattari, Felix, *A Thousand Plateaus: Capitalism & Schizophrenia* trans. by Brian Massumi (Minneapolis, MN: University of Minnesota Press, 1987).

Derrida, Jacques, *Speech and Phenomena and Other Essays on Husserl's Theory of Signs* ed. and trans. by David Allison (Evanston, IL: Northwestern University Press, 1973).

———, *Of Grammatology* trans. by Gayatri Chakravorty Spivak (Baltimore, MD: Johns Hopkins University Press, 1976).

———, 'Freud and the Scene of Writing', in *Writing and Difference* trans. by Alan Bass (London: Routledge and Kegan Paul, 1978), pp. 196–231.

———, *Edmund Husserl's* Origin of Geometry: *An Introduction* trans. by John P. Leavey, Jr (Lincoln, NE: University of Nebraska Press, 1978).

———, *Positions* trans. by Alan Bass (Chicago, IL: University of Chicago Press, 1981).

———, 'Plato's Pharmacy', in *Dissemination* trans. by Barbara Johnson (Chicago, IL: University of Chicago Press, 1982), pp. 67–186.

———, 'Différance' trans. by Alan Bass, in *Margins – Of Philosophy* (Chicago, IL: University of Chicago Press, 1982), pp. 1–28.

———, 'The Ends of Man' trans. by Alan Bass, in *Margins – Of Philosophy* (Chicago, IL: University of Chicago Press, 1982), pp. 109–36.

———, 'The Principle of Reason: The University in the Eyes of Its Pupils' trans. by Catherine Porter and Edward P. Morris, *Diacritics* (Fall, 1983), pp. 6–20.

———, *Memoires: For Paul de Man* trans. by Cecile Lindsay, Jonathan Culler and Eduardo Cadava (New York: Columbia University Press, 1986).

———, 'Le main de Heidegger (*Geschlecht* II)', in *Psyché: Inventions de l'autre* (Paris: Galilée, 1987), pp. 415–71.

———, 'The Original Discussion of Différance', in *Derrida and Différance* ed. by Robert Bernasconi and David Wood (Evanston, IL: Northwestern University Press, 1988), pp. 83–97.

———, *Of Spirit: Heidegger and the Question* trans. by Geoffrey Bennington and Rachel Bowlby (Chicago, IL: University of Chicago Press, 1989).

———, 'Some Statements and Truisms about Neo-logisms, Newisms, Postisms, Parasitisms and Other Small Seisisms' trans. by A. Tomiche, in *The States of 'Theory': History, Art and Critical Discourses* ed. by David Carroll (New York: Columbia University Press, 1991), pp. 63–94.

———, 'Politics and Friendship: An Interview with Jacques Derrida' trans. by Robert Harvey, in *The Althusserian Legacy* ed. by E. Ann Kaplan and Michael Sprinker (London: Verso, 1993), pp. 183–231.

———, *Specters of Marx: The State of the Debt, the Work of Mourning and the New International* trans. by Peggy Kamuf (New York: Routledge, 1994).

———, 'The Rhetoric of Drugs', in *Points . . . Interviews, 1974–1994* ed. by Elizabeth Weber, and trans. by Peggy Kamuf (Stanford, CA: Stanford University Press, 1995), pp. 228–54.

———, *The Gift of Death* trans. by David Wills (Chicago, IL: University of Chicago Press, 1995).

———, *Archive Fever: A Freudian Impression* trans. by Eric Prenowitz (Chicago, IL: University of Chicago Press, 1996).

———, 'Faith and Knowledge: The Two Sources of "Religion" at the Limits of Reason Alone' trans. by Samuel Weber, in *Religion* ed. by Jacques Derrida and Gianni Vattimo (Stanford, CA: Stanford University Press, 1998), pp. 1–78.

———, 'Typewriter Ribbon: Limited Ink (2)', in *Material Events: Paul de Man and the Afterlife of Theory* ed. by Tom Cohen, Barbara Cohen, J. Hillis Miller and Andrzej Warminski (Minneapolis, MN: University of Minnesota Press, 2001), pp. 277–360.

———, 'Nietzsche and the Machine' trans. by Richard Beardsworth, in *Negotiations: Interventions and Interviews 1971–2000* ed. by Elizabeth Rottenberg (Stanford, CA: Stanford University Press, 2002), pp. 215–56.

———, *On Touching – Jean-Luc Nancy* trans. by Christine Irizarry (Stanford, CA: Stanford University Press, 2005).

———, *Paper Machine* trans. by Rachel Bowlby (Stanford, CA: Stanford University Press, 2005).

——— and Stiegler, Bernard, *Echographies of Television: Filmed Interviews* trans. by Jennifer Bajorek (Cambridge: Polity, 2002).

———, Sous la direction de Marie-Louise Mallet, *L'animal autobiographique: Autour de Jacques Derrida* (Paris: Galilée, 1999).

De Saussure, Ferdinand, *Course in General Linguistics* trans. by Wade Baskin (New York: McGraw-Hill, 1959; 2nd edition, 1966).

Descartes, René, *Treatise on Man* trans. by Thomas Steele Hall (Cambridge, MA: Harvard University Press, 1972).

———, *The Philosophical Writings of Descartes* trans. by John Cottingham, Robert Stoothoff and Dugald Murdoch, volume 1 (Cambridge: Cambridge University Press, 1985).

———, *Meditations on First Philosophy* trans. by John Cottingham (Cambridge: Cambridge University Press, 1986).

De Vries, Hent, *Philosophy and the Turn to Religion* (Baltimore, MD: Johns Hopkins University Press, 1999).

Dillon, Michael and Reid, Julian, *The Liberal Way of War: Killing to Make Life Live* (London: Routledge, 2008).

Dolar, Mladen, *'La femme-machine'*, *New Formations* 23 (1994), pp. 43–54.

Dougherty, Stephen, 'New Media's Body', in *Language Systems: After Prague Structuralism* ed. by Louis Armand (Prague: Literaria Pragensia), pp. 65–78.

Ellman, Maud, 'Deconstruction and Psychoanalysis', in *Deconstructions: A User's Guide* ed. by Nicholas Royle (London: Palgrave Macmillan, 2000), pp. 211–37.

Engels, Friedrich, 'The Part Played by Labour in the Transition from Ape to Man', in *The Dialectics of Nature* trans. by Clemens Dutt (London: Lawrence & Wishart, 1940), pp. 270–97.

Espinas, Alfred, 'L'Organisation ou la machine vivante en Grèce au IVe siècle avant J-C', *Revue du métaphysique et de morale* (1903), pp. 702–15.

Everett, Hugh, 'The Theory of the Universal Wavefunction', in *The Many-Worlds Interpretation of Quantum Mechanics* ed. by Bryce DeWitt and R. Neill Graham, Princeton Series in Physics (Princeton, NJ: Princeton University Press, 1973), pp. 3–140.

Feenberg, Andrew, 'Technology and Meaning', in *Questioning Technology* (New York and London: Routledge, 1999), pp. 183–99.

Forrester, John, *The Seductions of Psychoanalysis: Freud, Lacan and Derrida* (Cambridge: Cambridge University Press, 1990).

Foucault, Michel, *Security, Territory, Population: Lectures at the Collège de France 1977–78* trans. by Graham Burchell (Basingstoke and New York: Palgrave Macmillan, 2007).

———, *Birth of Biopolitics: Lectures at the Collège de France 1979–80* trans. by Graham Burchell (Basingstoke and New York: Palgrave Macmillan, 2008).

Frank, Manfred, *Was ist Neostructuralismus?* (Frankfurt am Main: Suhrkamp, 1983).

Freud, Sigmund, 'The Uncanny', in *The Complete Psychological Works*, volume 17 trans. by James Strachey (London: Hogarth Press, 1953–74), pp. 217–52.

———, 'A Note upon the "Mystic Writing Pad" ', in *The Complete Psychological Works*, volume 19 trans. by James Strachey (London: Hogarth Press, 1953–74), pp. 227–32.

———, 'The Ego and the Id', in *The Complete Psychological Works*, volume 19 trans. by James Strachey (London: Hogarth Press, 1953–74).

———, *The Origins of Psychoanalysis: Letters to Wilhelm Fleiss, Drafts and Notes, 1887–1902* ed. by M. Bonaparte, A. Freud and E. Kris (New York: Basic Books, 1954).

———, 'Project for a Scientific Psychology', in *The Origins of Psychoanalysis: Letters to Wilhelm Fleiss, Drafts and Notes, 1887–1902* ed. by M. Bonaparte, A. Freud and E. Kris (New York: Basic Books, 1954), pp. 347–445.

———, *Beyond the Pleasure Principle* trans. by James Strachey (New York: W. W. Norton, 1961).

Funkenstein, Amos, *Theology and the Scientific Imagination from the Middle Ages to the Seventeenth Century* (Princeton, NJ: Princeton University Press, 1986).

Graham, Elaine L., *Representations of the Post/Human: Monsters, Aliens and Others in Popular Culture* (Manchester: Manchester University Press, 2002).

Hägglund, Martin, *Radical Atheism: Derrida and the Time of Life* (Stanford, CA: Stanford University Press, 2008).

Hansen, Mark B. N., *Embodying* Technesis: *Technology Beyond Writing* (Ann Arbor, MI: University of Michigan Press, 2000).

———, 'The Time of Affect, or Bearing Witness to Life', *Critical Inquiry* 30: 3 (2004), pp. 584–626.

———, 'Realtime Synthesis and the Différance of the Body: Technocultural Studies in the Wake of Deconstruction', *Cultural Machine* 6 (2004) (http://www.culturemachine.net/index. php/cm/article/view/9/8), accessed 1 October 2009.

———, *New Philosophy for New Media* (Cambridge, MA: MIT Press, 2004).

———, *Bodies in Code: Interfaces with Digital Media* (New York: Routledge, 2006).

———, 'Media Theory', *Theory, Culture & Society* 23: 2–3 (2006), pp. 297–306.

Haraway, Donna, 'A Cyborg Manifesto: Science, Technology and Socialist-Feminism in the Late Twentieth Century', in *Simians, Cyborgs and Women: The Reinvention of Nature* (New York: Routledge, 1991), pp. 149–81.

Hardt, Michael and Negri, Antonio, *Empire* (Cambridge, MA and London: Harvard University Press, 2000).

Harman, Graham, *Tool-Being: Heidegger and the Metaphysics of Objects* (New York: Open Court, 2002).

Hart, Kevin, *The Trespass of the Sign: Mysticism and Deconstruction* (Cambridge: Cambridge University Press, 1989).

Hayles, N. Katherine, *How We Became Posthuman: Virtual Bodies in Cybernetics, Literature and Informatics* (Chicago, IL and London: University of Chicago Press, 1999).

Hegel, G. W. F., *Hegel's Philosophy of Nature* trans. by M. J. Perry (London: Allan & Unwin, 1970).

Heidegger, Martin, *Being and Time* trans. by John Macquarrie and Edward Robinson (Oxford: Blackwell, 1962).
———, 'The Origin of the Work of Art' trans. by Albert Hofstadter, in *Poetry Language Thought* (New York: HarperCollins: 1971), pp. 15–86.
———, 'The Question Concerning Technology', in *The Question Concerning Technology and Other Essays* trans. by William Lovitt (New York: Harper & Row, 1977), pp. 3–35.
———, 'Science and Reflection', in *The Question Concerning Technology and Other Essays* trans. by William Lovitt (New York: Harper & Row, 1977), pp. 155–82.
———, 'Letter on Humanism' trans. by Frank A. Capuzzi, in *Basic Writings* (London: Routledge, 1978), pp. 217–65.
———, *The Basic Problems of Phenomenology* trans. by Albert Hofstadter (Bloomington, IN: Indiana University Press, 1982).
———, 'Martin Heidegger in Conversation with Richard Wisser' trans. by Lisa Harries, in *Martin Heidegger and National Socialism* ed. by Günther Neske and Emil Kettering (New York: Paragon House, 1990), pp. 81–8.
———, *Basic Questions of Philosophy: Selected 'Problems' of 'Logic'* trans. by Richard Rojcewicz and André Schuwer (Bloomington, IN: Indiana University Press, 1994).
———, 'On the Essence and Concept of *φύσις* in Aristotle's *Physics* B, I' trans. by Thomas Sheehan, in *Pathmarks* ed. by William McNeil (Cambridge: Cambridge University Press, 1998), pp. 183–230.
———, 'Plato's Doctrine of Truth' trans. by Thomas Sheehan, in *Pathmarks* ed. by William McNeil (Cambridge: Cambridge University Press, 1998), pp. 155–82.
———, *Contributions to Philosophy (From Enowning)* trans. by Parvis Emad and Kenneth Maly (Bloomington, IN: Indiana University Press, 1999).
———, *Introduction to Metaphysics* trans. by Gregory Fried and Richard Polt (New Haven, CT and London: Yale University Press, 2000).
———, *Zollikon Seminars: Protocols – Conversations–Letters* ed. by Medard Boss and trans. by Franz Mayr and Richard Askay (Evanston, IL: Northwestern University Press, 2001).
Heisenberg, Werner, 'Über quantentheoretische Umdeutung kinematischer und mechanischer Beziehungen', *Zeitschrift für Physik* 33 (1927), pp. 879–93.
———, 'Das Naturbild der heutigen Physik', in *Die Künste im techischen Zeitalter* (Munich: Oldenbourg, 1954), pp. 43–69.
Heller, Agnes, 'Paradigm of Production: Paradigm of Work', *Dialectical Anthropology* 6 (1981), pp. 71–9.
Heller-Roazen, Daniel, *The Inner Touch: Archaeology of a Sensation* (Cambridge, MA: MIT Press and Zone Books, 2007).
Hemming, Laurence Paul, *Heidegger's Atheism: The Refusal of a Theological Voice* (Indiana, IN: University of Notre Dame Press, 2002).
Herz, Neil, 'Freud and the Sandman', in *The End of the Line: Essays on Psychoanalysis and the Sublime* (New York: Columbia University Press, 1985), pp. 97–121.
Hodge, Joanna, *Derrida on Time* (London: Routledge, 2007).
Hoffman, E. T. A., *Tales of E. T. A Hoffman* trans. by Leonard J. Kent and Elizabeth C. Knight (Chicago, IL: University of Chicago Press, 1969).
Hunter, Ian, *Rival Enlightenments: Civil and Metaphysical Philosophy in Early Modern Germany* (Cambridge: Cambridge University Press, 2001).

Husserl, Edmund, *The Phenomenology of Internal Time Consciousness* trans. by J. Churchill (Bloomington, IN: Indiana University Press, 1964).

Irigaray, Luce, *Speculum of the Other Woman* trans. by Gillian C. Gill (New York: Columbia University Press, 1985).

Jacob, François, *La logique du vivant: Une histoire de l'héredité* (Paris: Gallimard, 1970).

Janicaud, Dominique, Courtine, Jean-François, Chrétien, Jean-Louis, Henry, Michel, Marion, Jean-Luc and Ricoeur, Paul, *Le tournant théologique de la phénoménologie française* (Paris: L'Eclat, 1991).

Jentsch, Ernst, 'On the Psychology of the Uncanny' trans. by Roy Sellars, *Angelaki* 2: 1 (1995), pp. 7–16.

Johnson, Christopher, *System and Writing in the Philosophy of Jacques Derrida* (Cambridge: Cambridge University Press, 1993).

———, 'I-You-We, Robot', in *Technicity* ed. by Arthur Bradley and Louis Armand (Prague: Charles University Press, 2006), pp. 101–28.

———, 'Derrida and Technology', in *Derrida's Legacies: Literature and Philosophy* ed. by Robert Eaglestone and Simon Glendinning (London: Routledge, 2008), pp. 54–65.

Joyce, James, *Portrait of the Artist as a Young Man* (London: Penguin, 2000).

Kant, Immanuel, *Critique of Judgement* trans. by Werner S. Pluhar (Indianapolis, IN and Cambridge: Hackett, 1987).

———, 'A Renewed Attempt to Answer the Question "Is the Human Race Continually Improving?" ', in *Political Writings* ed. and intr. by Hans Reiss and trans. by H. B. Nisbet (Cambridge: Cambridge University Press, 2nd edition, 1991), pp. 176–90.

Kittler, Friedrich, *Discourse Networks 1800–1900* trans. by Michael Metteer with Chris Collins (Stanford, CA: Stanford University Press, 1995).

———, 'The World of the Symbolic – A World of the Machine', in *Literature, Media, Information Systems* ed. and intr. by John Johnston (Amsterdam: OPA, 1997), pp. 130–46.

———, *Gramophone, Film, Typewriter* trans. by Geoffrey Winthrop-Young and Michael Wutz (Stanford, CA: Stanford University Press, 1999).

Kofman, Sarah, *Freud and Fiction* trans. by Sarah Wykes (Boston, MA: Northwestern University Press, 1991).

Kristeva, Julia, *Revolution in Poetic Language* trans. by Margaret Waller (New York: Columbia University Press, 1984).

Kurzweil, Ray, *The Singularity is Near* (London: Duckworth, 2006).

Lacan, Jacques, *The Four Fundamental Concepts of Psychoanalysis* trans. by Alan Sheridan (Harmondsworth: Penguin, 1977).

———, *The Seminar of Jacques Lacan, Book I: Freud on Technique* trans. by John Forrester and ed. by Jacques-Alain Miller (Cambridge: Cambridge University Press, 1988).

———, *The Seminar of Jacques Lacan: Book II: The Ego in Freud's Theory and in the Technique of Psychoanalysis 1954–5* trans. by Sylvana Tomaselli and ed. by Jacques-Alain Miller (Cambridge: Cambridge University Press, 1988).

———, *The Seminar of Jacques Lacan Book III: The Psychoses* trans. by Russell Grigg (London: Routledge, 1993).

Leroi-Gouhran, André, *Gesture and Speech* trans. by A. Bostock Berger (Cambridge, MA: MIT Press, 1993).

Levinas, Emmanuel, *Totality and Infinity: An Essay on Exteriority* trans. by Alphonso Lingis (Pittsburgh, PA: Duquesne University Press, 1969).

Liu, Yu, 'The Beautiful and the Sublime: Kant's Paradise Lost and Paradise Regained', *Studies in Romanticism* 42: 3 (2003), pp. 187–202.

Lyotard, Jean-François, 'Kant Notice 4', in *The Differend: Phrases in Dispute* trans. by Georges Van Der Abbeele (Minneapolis, MN: University of Minnesota Press, 1988), pp. 161–69.

———, *The Inhuman: Reflections on Time* trans. by Geoffrey Bennington and Rachel Bowlby (Cambridge: Polity, 1991).

———, *Lessons on the Analytic of the Sublime* trans. by Elizabeth Rottenberg (Stanford, CA: Stanford University Press, 1994).

Malabou, Catherine, *What Should We Do with Our Brain?* trans. by Sebastian Rand (New York: Fordham University Press, 2008).

Marx, Karl, *Capital*, volume 3 (London: Lawrence & Wishart, 1970).

———, *The Economic and Philosophic Manuscripts of 1844* (London: Lawrence & Wishart, 1973).

———, *Grundrisse: Foundations of the Critique of Political Economy* trans. by Martin Nicolaus (London: Penguin, 1973).

———, *Selected Writings* ed. by D. McLellan (Oxford: Oxford University Press, 1977).

———, *Capital*, volume 1 trans. by Ben Fowkes (Harmondsworth: Penguin, 1990).

——— and Engels, Friedrich, *The German Ideology* (London: Lawrence & Wishart, 1970).

Maturana, Humberto R. and Valera, Francisco J., *Autopoiesis and Cognition: The Realisation of the Living* (Dordrecht: D. Reidel, 1980).

McGrew, W. C., *Chimpanzee Material Culture: Implications for Human Evolution* (Cambridge: Cambridge University Press, 1992).

Meillassoux, Quentin, *After Finitude: An Essay on the Necessity of Contingency* trans. by Ray Brassier (London: Continuum, 2008).

Merleau-Ponty, Maurice, *Phenomenology of Perception* trans. by Colin Smith (New York: Humanities Press, 1962).

Milesi, Laurent, 'Almost Nothing at the Beginning: The Technicity of the Trace in Deconstruction', in *Language Systems: After Prague Structuralism* ed. by Louis Armand and Pavel Černovsky (Prague: Charles University Press, 2007), pp. 22–41.

Miller, Jacques-Alain, 'An Introduction to Seminars I and II: Lacan's Orientation Prior to 1953', in *Reading Seminars I and II: Lacan's Return to Freud* ed. by Richard Feldstein, Bruce Fink and Maire Jannus (New York: SUNY, 1996), pp. 3–38.

Minsky, Marvin, *The Society of Mind* (London: Simon and Schuster, 1987).

Monod, Jacques, *Chance and Necessity, On the Natural Philosophy of Modern Biology* (Harmondsworth: Penguin, [1970] 1997).

Moravec, Hans, *Robot: Mere Machine to Transcendent Mind* (Oxford: Oxford University Press, 1999).

Nancy, Jean-Luc, *Corpus* (Paris: Editions Metailié, 2000).

———, *A Finite Thinking* trans. by Simon Sparks (Stanford, CA: Stanford University Press, 2003).

———, *La Déclosion: Déconstruction du christianisme 1* (Paris: Galilée, 2005).

Negri, Antonio, *Marx Beyond Marx: Lessons on the* Grundrisse trans. by Harry Cleaver, Michael Ryan and Maurizzio Viano (New York: Autonomedia/Pluto, 1991).

Osborne, Peter, 'Marx and the Philosophy of Time', *Radical Philosophy* 147 (2008), pp. 15–22.
Pearson, Keith Ansell, *Viroid Life: Perspectives on the Transhuman Condition* (London: Routledge, 1997).
Plato, *'Protagoras' and 'Meno'* trans. by R. C. Bartlett (Ithaca, NY: Cornell University Press, 2004).
Plotnitsky, Arkady, *Complementarity: Anti-Epistemology after Bohr and Derrida* (Durham, NC: Duke University Press, 1994).
Postone, Moishe, *Time, Labour and Social Domination: A Reinterpretation of Marx's Critical Theory* (Cambridge: Cambridge University Press, 1993).
Protevi, John, *Political Physics* (London: Continuum, 2001).
Rabinbach, Anson, *The Human Motor: Energy, Fatigue and the Origins of Modernity* (Berkeley, CA: University of California Press, 1992).
Rojcewicz, Richard, *The Gods and Technology: A Reading of Heidegger* (Albany: SUNY, 2006).
Ronnell, Avital, *The Telephone Book: Technology – Schizophrenia – Electric Speech* (Lincoln, NE and London: University of Nebraska Press, 1989).
Rose, Nikolas, *The Politics of Life Itself: Biomedicine, Power, and Subjectivity in the Twenty-First Century* (Princeton, NJ: Princeton University Press, 2006).
Royle, Nicholas, *The Uncanny* (Manchester: Manchester University Press, 2003).
Salanskis, Jean-Michel, 'Ecce Faber', *Les Temps Modernes* 55: 608 (2000), pp. 235–78.
Sample, Ian, 'Craig Ventner Creates Synthetic Life Form', *Guardian*, 20 May 2010.
Santner, Eric L., *On Creaturely Life: Rilke/Benjamin/Sebald* (Chicago, IL: University of Chicago Press, 2006).
Scharff, Robert C. and Dusek, Val eds, *The Philosophy of Technology: The Technological Condition*, Blackwell Philosophy Anthologies (Oxford: Blackwell, 2003).
Schmidt, Alfred, *The Concept of Nature in Marx* trans. by Ben Fowkes (London: New Left Books, 1970).
Schoenbohm, Susan, 'Heidegger's Interpretation of *Phusis*', in Heidegger's *Introduction to Metaphysics* ed. by Richard Polt and Gregory Fried (New Haven, CT and London: Yale University Press, 2000), pp. 143–60.
Simondon, Gilbert, *Du mode d'existence des objets techniques* (Paris: Aubier, 1958).
———, *L'individuation psychique et collective* (Paris, Aubier, 1989).
Smith, Joseph H. and Kerrigan, William eds, *Taking Chances: Derrida, Psychoanalysis and Literature* (Baltimore, MD: Johns Hopkins University Press, 1984).
Stiegler, Bernard, *La Technique et le temps 2: La Désorientation* (Paris: Galilée, 1996).
———, *Technics and Time 1: The Fault of Epimetheus* trans. by Richard Beardsworth and George Collins (Stanford, CA: Stanford University Press, [1994] 1998).
———, 'Leroi-Gouhran: L'Inorganique organisé', *Cahiers de Médiologie* 6, 2e Semestre (1998) (http://www.mediologie.org/collection/06_mediologues/stiegler.pdf), accessed 1 October 2009.
———, *La Technique et le temps 3: Le Temps du cinema et la question du mal-etre* (Paris: Galilée, 2001).
———, 'Derrida and Technology: Fidelity at the Limits of Deconstruction and the Prosthesis of Faith' trans. by Richard Beardsworth, in *Jacques Derrida and the Future of the Humanities* ed. by Tom Conley (Cambridge: Cambridge University Press, 2002), pp. 238–70.
———, *De la misère symbolique 1: L'époque hyperindustrielle* (Paris: Galilée, 2004).

———, *Philosopher par accident: Entretiens avec Elie During* (Paris: Galilée, 2004).
———, '*Anamnēsis* & *Hypomnēsis*: The Memories of Desire' trans. by Arthur Bradley and François-Xavier Gleyzon, in *Technicity* ed. by Arthur Bradley and Louis Armand (Prague: Charles University Press, 2006), pp. 15–41.
———, *Des Pieds et des mains* (Paris: Bayard Culture, 2006).
———, *Economie de l'hypermateriel et psychopouvoir: Entretiens avec Philippe Petit et Vincent Bontems* (Paris: Mille et une nuits, 2006).
———, *Prendre soin: De la jeunesse et des générations* (Paris: Flammarion, 2008).
Taylor, Charles, *A Secular Age* (Cambridge, MA and London: Harvard University Press, 2007).
Thacker, Eugene, *Biomedia* (Minneapolis, MN: University of Minnesota Press, 2004).
Thurtle, Phillip and Mitchell, Robert eds, *Data Made Flesh: Embodying Information* (New York: Routledge, 2004).
Turing, Alan, 'Computing Machinery and Intelligence', *Mind* LIX: 236 (1950), pp. 430–60.
Valera, Francisco and Bourgine, Paul eds, *Toward a Practice of Autonomous Systems: Proceedings of the First European Conference on Artificial Life* (Cambridge, MA: MIT Press, 1992).
Weiner, Norbert, *Cybernetics or Control and Communication in the Animal and the Machine* (Cambridge, MA: MIT Press, 1948).
Wendling, Amy E., *Marx on Technology and Alienation* (New York: Palgrave Macmillan, 2009).
Williams, James, *Lyotard and the Political* (London: Routledge, 2000).
Wills, David, *Prosthesis* (Stanford, CA: Stanford University Press, 1995).
———, *Dorsality: Thinking Back Through Technology and Politics* (Minneapolis, MN and London: University of Minnesota Press, 2008).
Wolfe, Cary ed., *Zoontologies: The Question of the Animal* (Minneapolis, MN: Minnesota University Press, 2003).
Wolfe, Cary, *What is Posthumanism?* (Minneapolis, MN and London: University of Minnesota Press, 2010).
Woolfson, Charles, *The Labour Theory of Culture: A Re-Examination of Engels's Theory of Human Origins* (London: Routledge and Kegan Paul, 1982).
Zimmerman, Michael E., *Heidegger's Confrontation with Modernity: Technology, Politics, Art* (Bloomington, IN: Indiana University Press, 1990).
Žižek, Slavoj, *The Indivisible Remainder* (London: Verso, 1996).
———, *On Belief* (London: Routledge, 2001).

Index

www.ingramcontent.com/pod-product-compliance
Lightning Source LLC
Chambersburg PA
CBHW071644030826
49219CB00005B/349
9780230576926